Missing Bodies

BIOPOLITICS

MEDICINE, TECHNOSCIENCE, AND HEALTH IN THE 21ST CENTURY

General Editors: Monica J. Casper and Lisa Jean Moore

Missing Bodies: The Politics of Visibility
Monica J. Casper and Lisa Jean Moore

Missing Bodies

The Politics of Visibility

Monica J. Casper
and Lisa Jean Moore

NEW YORK UNIVERSITY PRESS
New York and London

NEW YORK UNIVERSITY PRESS
New York and London
www.nyupress.org

© 2009 by New York University

Library of Congress Cataloging-in-Publication Data

Casper, Monica J., 1966–
Missing bodies : the politics of visibility /
Monica J. Casper and Lisa Jean Moore.
p. cm. — (Biopolitics : medicine, technoscience,
and health in the 21st Century)
Includes bibliographical references and index.
ISBN-13: 978–0–8147–1677–9 (cl : alk. paper)
ISBN-10: 0–8147–1677–6 (cl : alk. paper)
ISBN-13: 978–0–8147–1678–6 (pb : alk. paper)
ISBN-10: 0–8147–1678–4 (pb : alk. paper)
1. Body, Human (Philosophy) 2. Body image. 3. Mortality.
4. Masculinity. 5. Equality. I. Moore, Lisa Jean, 1967– II. Title.
B105.B64.C36   2009
306.4'613—dc22   2009000609

New York University Press books are printed on acid-free paper, and
their binding materials are chosen for strength and durability. We
strive to use environmentally responsible suppliers and materials to
the greatest extent possible in publishing our books.

Manufactured in the United States of America
c   10 9 8 7 6 5 4 3 2 1
p   10 9 8 7 6 5 4 3 2 1

For a cherished mentor, Virginia L. Olesen
and
In memory of HM2 Charles Luke Milam (1981–2007)

Contents

Acknowledgments

This book is the product of almost two decades of conversations about health, medicine, bodies, relationships, sexuality, mothering, sociology, gender studies—indeed, about life itself. Since our early days in graduate school at the University of California, San Francisco, beginning in 1991, our individual scholarship has been fostered, prodded, sustained, challenged, and otherwise shaped by our various collaborations with each other. We are deeply appreciative of our long friendship.

We are grateful to the following institutions and individuals for scholarly support and encouragement:

For being an incomparable, dedicated, and brilliant editor, Ilene Kalish.

For careful and detailed editorial feedback on the entire manuscript, Paisley Currah, Grace Mitchell, and an anonymous reviewer for NYU Press.

For research assistance, Patricia Conway, Dan Morrison, Harmony Newman, Haley Swenson, and Heather Laine Talley.

For feedback on specific sections or chapters, Laura Carpenter, Lenora Champagne, Kim Christensen, Tim Fann, Liz Loeb, Laura Mamo, Shaka McGlotten, Joe Rollins, Heather Laine Talley, Rachel Washburn, Vanderbilt doctoral students in "Biopolitics and Biopower," and Purchase College undergraduates in "Birth and Death" and "Men and Masculinities."

For helping to keep things running during a very difficult year, the fabulous folks in Women's and Gender Studies at Vanderbilt University, especially Gayle Parrott and Shubhra Sharma.

For financial resources, Vanderbilt University, particularly the College of Arts and Science, and Arizona State University.

For collegial support and assistance, Suzanne Kessler, Mary Kosut, and Christina Williams at Purchase College and Lynn Clinton-Van Patter and Mary E. Bauer at Arizona State University.

For travel funds, the Purchase College Foundation, State University of New York.

And for helpful editorial assistance, especially with our images, Aiden Amos at NYU Press.

Sometimes during the writing of this book, our own bodies were missing in action, and we needed other bodies to step in for us or to patiently await our return to "real life."

Lisa thanks her husband Paisley Currah for his unfailing love, support, and faith; Grace and Georgia Moore for being amazing and sparkling daughters; Mira Handman for pillow talk; and Patti Curtis and Shari Colburn for their heartwarming acceptance and friendship through life's challenges.

Monica thanks her daughters, Mason and Delaney, for their patience and wonderful distractions; Paige Clayton for providing loving and consistent childcare; Patricia and Dennis Struck for being groovy parents; Tanya Casper for her unstinting loyalty; Adele Clarke for inspiration and friendship; and Heather Talley, Dana Nelson, and Asante Todd for making one of life's tough spots more bearable.

We thank the Milam family, especially Ri and Michael, for graciously allowing us to talk about Luke's death in our book.

Inspection functions ceaselessly. The gaze is alert everywhere . . .
                              Michel Foucault, *Discipline and Punish* (1977)

What does it mean for a country to choose blindness as its national pledge of allegiance?
                              Avery Gordon, *Ghostly Matters* (1996)

The right speculum for the job makes visible the data structures that are our bodies.
                    Donna Haraway, *Modest_Witness@Second_Millennium* (1997)

Let the atrocious images haunt us. . . . The images say: This is what human beings are capable of doing—may volunteer to do, enthusiastically, self-righteously. Don't forget.
                    Susan Sontag, *Regarding the Pain of Others* (2004)

1

Introduction

The Bodies We See, and Some That Are Not Here

> Gloria liked the idea that there were cameras watching everyone
> everywhere. Last year Graham had installed a new state-of-the-art
> security system in the house—cameras and infrared sensors and
> panic buttons and goodness knows what else. Gloria was fond of the
> helpful little robots that patrolled her garden with their spying eyes.
> Once, the eye of God watched people, now it was the camera lens.
>
> Kate Atkinson, *One Good Turn* (2006)

We live in an age of proliferating human bodies, both literally
and figuratively. The world's population is more than six and a half *billion,*
a staggering number by any measure—and perhaps too many people for
one fragile, embattled planet and current allocations of resources. Representations of these omnipresent, multiplying bodies are both enhanced
and amplified via new biomedical, digital, and representational technologies, like MRIs and sonograms. Bodies are made visible and seen—or
watched, to embrace the conspiratorial—via a range of globalized practices.[1] Indeed, the human body has never been more *visible* and rapidly
mobile (and mobilized) than it is in the first decade of the 21st century.
It should not surprise anyone that the U.S. National Library of Medicine
sponsors an ambitious digital image library of adult anatomy called the
Visible Human Project or that the exhibit Body Worlds, featuring plas-
tinated, posed human cadavers, has been both wildly successful and intensely controversial.

From a very young age, human beings are trained to visually process
and meticulously read bodies—our own and others—for social cues about
love, beauty, status, and identity. Bodies are socially constructed within

social orders, including patterns of dominance and submission along lines of race, ethnicity, gender, age, and physiological normativity. Accurately reading the body of another, beginning with our mothers and other care-givers immediately after birth, can sometimes mean the difference between survival and death. As such, the visualized body is powerfully symbolic in a multitude of ways and across often quite-contested domains. Increasingly, too, our bodies are under surveillance, digitized and processed for analysis. In a suspicious post-9/11 world, a marked hysteria accompanies the quest for visual proof of human beings' whereabouts, activities, interactions, purchases, conversations, and migrations.

Cameras perch on lampposts and rooftops in towns and cities across the United States and in other countries, monitoring all sorts of public interactions, movements, and activities. Community activists in New York City, including the Surveillance Camera Players, estimate that there are at least 10,000 security cameras in Manhattan alone, while a European group estimates that London has some 500,000 public cameras. The Scripps Howard News Service reports there are about 5 million video surveillance cameras in use in the United States today and that the number of government-funded cameras has grown exponentially, courtesy of dollars earmarked for "homeland security" needs. The security and private protection industry is worth about $9 billion annually and is expected to grow to a $20 billion industry by 2010.

Recent congressional debate in the United States centered on H.R. 418, the Real ID Act, which mandates the creation of a national identification card with multiple data storage points. In conjunction with the Departments of Motor Vehicles, all states will be required to comply with provisions of the act by December 31, 2009. These may include the Combined DNA Indexing System (CODIS) of the FBI, Medicare and Medicaid records, military records, criminal records, immigration status, employment information, credit reports, and so on. The Real ID Act and other government and private surveillance efforts, coupled with the explosion of visual and biometric technologies, are making human bodies partially and wholly legible with limited to no public discussion and shockingly little regulation. In the name of national security, it seems, the neoliberal state is watching all of us—and it is marking the bodies of citizens and especially noncitizens at an unprecedented level and with as yet largely unknown consequences.

This surveillance by our paternal uncle, let's call him Sam, and his cronies is taking place in a culture that seems obsessed with, and made almost frantic by, "real" bodies doing "real" things in "real" time. It sometimes seems as if we are more attuned to the daily activities of America's Top Model, Paris Hilton, college football players, and al-Qaeda than we are to those of our own teenage children or our neighbors in gated communities whose names we may not even know. And yet, despite this escalating Orwellian practice by which the government and corporations visually locate and define bodies so as to regulate (and perhaps punish) them, some bodies are conspicuously missing in action.

Not all bodies are equally visible to the cameras, the watchers, or the analysts; indeed, some bodies may not necessarily want to be seen at all. Nor do all of the discursive, visual, and geographic sites show the full panoply of human bodies that might be present. Certainly there are bodies that we see so routinely that they appear in our dreams; we ourselves may be watched regularly, skin crawling, at every turn. But there are some bodies that are invisible, that have disappeared, or whose absence is unaccounted for and not remarked on in popular culture or by government agencies such as, to name just one example, FEMA (Federal Emergency Management Agency).

In *Missing Bodies,* we are interested in exploring how certain places, spaces, policies, and practices in contemporary society, particularly in the United States, exhibit and celebrate some bodies while erasing and denying others. What can account for the fact that certain bodies are hyper-exposed, brightly visible, and magnified, while others are hidden, missing, and vanished? We believe there are dimensions of corporeal visibility and erasure that need to be charted and interpreted, for intellectual and political reasons, and we attempt to do so here. Interested in social processes and conditions of local and global stratification, or the many ways in which the world's people are unequal, we investigate in this book the traffic *between* and *among* visible, invisible, and missing bodies.

At the same time, we strategically deploy the multiple uses of the term "missing" to interrogate the ways in which we are *affectively* missing certain bodies. For to be missing means that something or someone was once visible and is now lost. Thus "missing" is a kind of invisibility, one usually characterized by a high degree of emotion, as with missing children or soldiers M.I.A. (missing in action). As feminist sociologists of the body committed to ethical practices and social justice, we find ourselves longing for these missing bodies and for stories about them. For example, why did it take so long for the U.S. media to begin telling stories about the "falling bodies" of 9/11, those tragic figures who leaped from the burning, crumbling towers rather than be obliterated by flames (Flynn and Dwyer 2004)? Why has the U.S. government refused to allow photographs or filming of flag-draped coffins carrying the bodies of soldiers who are fighting and dying in Iraq and Afghanistan?

In addition, as mothers of young children—we each have two daughters—we have been outraged by the media erasure of women and children devastated by recent "natural" disasters along the American Gulf Coast and in Southeast Asia. Furthermore, we are incensed that even though the United Nations estimate for the Iraqi death toll exceeded 34,000 in 2006, there is almost complete lack of media coverage of, and government accounting for, the men, women, and children of all nationalities killed by American troops and their allies in Iraq or caught in the fatal crosshairs of ethnic and religious intolerance.[2] We wonder: Where are the "missing girls" of China and India and other patriarchal cultures, these young victims of "sex-selective" abortion and female infanticide? And we lament the massive incarceration of our own nation's young Black men, and increasingly Black women, hundreds of thousands of them warehoused behind bars and out of sight.

In short, because we care deeply about missing bodies in both a pragmatist and a humanist sense, we have turned our analytical lens to questions of corporeal presence and absence. Following scholars of the visible, whose work we highlight below, we suggest here that the visible and invisible dimensions of human life, including representations of bodies, work together to create social order as we know it. In this book, through a series of empirical case studies, we investigate the mechanisms by which some bodies can be found with varying degrees of ease in American popular culture, policy, and social theory, while others cannot. While seeking to develop new intellectual understandings of bodily visibility and erasure, we are also deeply committed to the redistribution of political and theoretical attention to missing bodies and to revealing the consequences of chronic inattention and inaction by scholars and others. This book, then, is a recovery project, forged with equal parts hope and fury.

Bodies in and of Social Theory

At the dawning of the 20th century, sociology emerged as a method of inquiry aimed at explaining the social causes and effects of seemingly personal acts. Importantly, sociology offered an alternative to the biological, psychological, and individualistic definitions of human action. If one conceives of the intellectual history of sociology as different strands of thought, one tendency is to treat the individual as a rational, disembodied, decision-making agent, a kind of talking head with no recognizable body. In many ways, this line of inquiry against the biologically determined notions of social order meant that corporeality—or the flesh, bone, functions, physiology, sensations, and materiality of the body—was for over a century ignored or merely taken for granted. But within the past three decades, social science, spurred by feminist theory and practice, has contributed robust analyses to academic explorations and explanations of the human body and its representations.

Because we are medical sociologists *and* feminist scholars, we work in a tradition of material social constructionism and thus ground our analysis in the historical sociocultural forces that have shaped and created bodies. We consider these bodies as shifting and plural, alive with multiple potentials. Conversely, essentialists believe that a pre-social natural body exists (Connell and Dowsett 1993), an idea that is anathema to our project here. Although corporeality must be acknowledged and integrated into social theory, corporeality itself is not static; it changes as our interpretations of it are modified over time (Clarke 1995), and it also changes in response to the physical world. Social scientists must resist the temptation to see corporeality as sui generis, even though bodies might appear to have obdurate and consistent physical characteristics. For at the same moment that actual physical bodies exist, our understandings of these bodies, our interpretations and explanations of bodily processes, give meaning to their materiality.

Sociologist Bryan Turner (1987, 1992) suggests four key social developments that contributed to growth in the sociological and cultural investigation of bodies:

1. The growth of consumer culture. Shifts in mass consumption from the 1920s led to the availability of cheap, durable goods, such as cosmetics, health aides, and fashion accessories, which helped to secularize the body into a vehicle of identity display. Further, Mike Featherstone, Mike Hepworth, and Bryan Turner's (1991) analysis of "the body in consumer culture" is helpful in understanding the Western cultural imperative of maintaining bodies in late capitalism. They argue, "self preservation depends upon the preservation of the body within a culture in which the body is the passport to all that is good in life. Health, youth, beauty, sex, fitness are the positive attributes which body care can achieve and preserve" (1991:186).

2. The development of postmodern themes in the arts, architecture, and the humanities. Postmodernist theorists problematize subject/object distinctions prevalent in modernist representations. The human body is not distinct from the self, they argue, but is deeply interrelated to identity and self-expression. Postmodern scholarship and methods of inquiry provide tools to read and "deconstruct" the human body.

3. The feminist movement. Although Turner cites the feminist movement in a broad sense, the women's health movement especially challenged predominant biomedical ways of constructing bodies (Ruzek 1978, Lewin and Olesen 1985). As both consumers and scholars, many women rebelled against the hegemonic medical establishment's strategies of medicalization and mystification of female bodily functions. These challenges to "thinking as usual" within medical settings encouraged many women to wage feminist critiques against the standardization of male bodies as the model for individualism and better health. As Moira Gatens (1992) argues, women are often forced to "elide" or suppress their own "corporeal specificity" to participate in liberal democracies.

4. The impact of Michel Foucault's scholarship in the social sciences and humanities. Foucault's work brought forth a rich anthropological, sociological, and historical analysis of the social production of individual bodies and populations through his understanding of discipline and surveillance. He argued that disciplinary power, focused on individuals, operates through institutions and discourses to make docile subjects and productive bodies. When these bodies are considered in aggregate—in other words, defined as a population—a new but related form of power emerges. In his genealogies, Foucault establishes biopower emerging at the beginning of the 19th century in the West. This power led to the proliferation of new regimes, each participating in the social production of distinctive populations: incarcerated bodies, homosexual bodies, insane bodies, reproductive bodies.

As Foucault argued, sovereign rule under a monarchy was displaced and replaced by democratic systems of rules and regulations. The juridical competition of experts (legal, psychoanalytic, medical), the development of discourses of rights, and the concerted effort of disciplines to standardize and normalize the body have together enabled the construction of modernist knowledge of and about human bodies. On a broader scale, biopolitics is defined as the social practices and institutions established to regulate a population's quality (and quantity) of life. Disciplinary power and biopower, which together can be understood as biopolitics, operate together to normalize individuals by coercing them, often by subtle mechanisms, to conform to standards and, in so doing, to create self-regulating pliant bodies and populations (e.g., Inda 2005, Rose 2007).

Further, Foucault uses the notion of the panopticon to illustrate the key role of surveillance and normalization in societies. Building on the work of 18th-century English philosopher Jeremy Bentham, Foucault expanded the theoretical scope of the panopticon outside of the prison industrial complex to everyday life. The metaphor of a circular prison, which enables an inspector to be an omnipotent functionary, explains how subjects learn to self-regulate their behavior. The panopticon model, when implemented in public settings, is virtually unnoticed. Briefly applied to contemporary life, we live in a panoptic society constantly inspected by regulatory agencies (like public health departments, the police, the fashion industry) that make the human body an object of the normalizing gaze. That is, bodies are objectified. And since we don't know when we are being watched, we learn to police ourselves.

Certainly, we are quick to note, we each have the potential to be meaning-making agents, so we can and do resist these forces of normalization. Many of us are often participating in or resisting health precautions, legal standards, and physical enhancements. Self-help health movements, such as HIV/AIDS activism and the women's health movement, provide just two examples of individuals resisting the dominant discourses of biomedicine (Clarke and Olesen 1999).

Inspired by Foucault and political theory, Judith Butler (1993) proposes a theory of materialization in order to confront the fear that certain social constructionists exhibit regarding physicality of bodies. She argues that it is *how* some bodies and parts of bodies come to matter that should be the focus of social constructionist analyses. Butler sees constructionism not as a one-shot fixed phenomenon but as "processes of reiteration": no singular power acts; rather, a persistent yet unstable repetition process itself is powerful. Matter is "a process of materialization that stabilizes over time to produce the effect of boundary, fixity and surface" (3). It is Butler's contention that constructionists, even the most radical, must at some point concede to the materiality of the body. Instead of beginning these inquiries from the point that bodies exist, Butler asserts that we must ask what are the regulatory norms through which bodies are materialized. How, in other words, are bodies erected?

One of Butler's most compelling arguments insists that agency must be reworked to avoid embracing the Enlightenment notions of voluntarism and free will while still retaining a theory of subversive performance. This free-willed agent is a regulatory myth. That is, to paraphrase Butler, the subject who resists these regulatory norms is also produced by these norms. Further, she argues that in order to understand the multiple forces of materialization, investigations should include looking at which bodies *fail to matter:* "How bodies which fail to materialize provide the necessary 'outside,' if not the necessary support, for the bodies which, in materializing the norm, qualify as bodies that matter" (Butler 1993:16).

Also highly influenced by Foucault, post-9/11 "security studies" is an emergent, interdisciplinary field that examines the intensifying of social control through the use of techniques of visibility directed at bodies and people's internalization of such control. State-sanctioned institutions refine techniques to see and observe the movements and behaviors of bodies through biometrics (methods for uniquely recognizing humans based on physical properties) and biotelemetrics (implantable devices for transmission of biological data). Surveillance cameras exercise extraordinary power by monitoring criminal activity, maintaining security, and controlling anything deemed to be deviant according to the ideology of those in power (Currah and Moore 2009). Although contemporary security and surveillance activities may be experienced as unprecedented, states have a long history of making bodies legible, both individually and in the aggregate. Social scientists Jane Caplan and John Torpey track "the nineteenth-century development of documentary practices through which every citizen, not just the delinquent or deviant, was to be made visible to the state: not by physical marks on the body, but by the indirect means of registrations, passes, censuses, and the like" (2001:8).

These emergent identification procedures drew on a repertoire of physical signs and measurements but represented them in written and visual records, both individually portable and centrally filed. Examples include birth certificates, passports, and medical records. Caplan and Torpey further argue:

> The elaboration of systematic regimes of representation disclosed a central tension in the project of identification, as opposed to mere classification. The identity document purports to be a record of uniqueness, but

also has to be an element in a classifying series that reduces individuality to a unit in a series, and that is thus simultaneously deindividualizing. This discloses the fundamental instability of the concept of the "individual" as such, and helps to explain the uneasy sense that we never fully own or control our identity, that the identity document carries a threat of expropriation at the same time as it claims to represent who we "are." (8)

As sociologist David Lyon has written, from modernity onward "the body achieved new prominence as a site of surveillance. Bodies could be rationally ordered through classification in order to socialize them within the emerging nation-state. Bodies were distrusted as sensual, irrational, and thus in need of taming, subject to disciplinary shaping toward new purposes. By associating a name, or later, a number with the body, each person could be distinguished from the next" (2001:292). One chilling example of this is concentration camp tattoos.

As methods of accounting for and watching bodies as they live, consume, get sick, and die have become more sophisticated and highly valuable to the corporate state, we believe that certain academic and applied fields of study have missed opportunities for critical intervention. (For an exception, see art historian Barbara Marie Stafford's 1991 masterpiece, *Body Criticism*.) As we discuss in the pages that follow, demography, epidemiology, and economics often lack critical grounding and instead reduce the understanding of human bodies and experiences to auditing operations. Establishing the rates, odds, ratios, and cost/benefit breakdown of bodies erases personhood and subjectivity in the name of the aggregate.

In short, social scientists and humanists have attended to bodies, producing a growing corpus of material on embodiment, embodied experiences, body regulation, bodywork, representations of bodies, and cultural exposures of the body. Yet there has been limited sociological or other attention to the visibility of bodies, including their deliberate erasure, their unanticipated disappearances and elisions, and their celebratory objectification. It is our contention here that bodies are omnipresent; as Butler argues, the materialization of bodies is part and parcel of the creation of social and political life. All discourses and practices rely on the actions, regulations, interactions, and positioning of human bodies and the agents inhabiting them. But because society is stratified along lines of gender, race, class, sexuality, age, disability status, citizenship, geography, and other cleavages, some bodies are public and visually dissected while others are vulnerable to erasure and marginalization.

A Sociology of That Which Is Not Always Observable

In *Missing Bodies,* we investigate several provocative discursive and visual sites in which bodies are essential to the shape and functioning of the site and are socially present but culturally are missing in action or assume a spectral form. Borrowing most obviously from Avery Gordon's (1996) astute and beautifully written analysis of "ghostly matters," we explore the ways in which the sites under discussion here are haunted by these translucent, potent bodies. We attend to the cultural politics at work in corporeal disappearance, as well as the social and economic consequences of visibility and invisibility as they relate to the privileges and benefits of citizenship. By bringing bodies back into the frame and making them *sociologically* visible, just as historians Robert Proctor and Londa Schiebinger (2008) have shown regarding "the making and unmaking of ignorance," we illuminate the complicated processes that render some bodies and relationships imperceptible to the naked eye. Yet we are also fully aware that we do not have the capacity to find all the bodies that need to be seen. We must acknowledge that there are some bodies that may be unknowable.

Gordon's eloquent work is about haunting as "a constituent element of modern social life." A sociology of haunting, in her view, is a method for pulling the "seething presence" of ghosts "out of the shadows" and into our analytic frame. To investigate haunting is to write a history of the present—sociology's special province, as Gordon names it, echoing Foucault. While acknowledging that ghosts are an odd topic for sociologists, she presents a compelling argument for why we should take them seriously. Drawing on three cases or "ghost stories"—the role of Sabrina Spielren in the history of psychoanalysis, the story of Argentina's disappeared as fictionalized by Luisa Valenzuela, and slavery in the United States as embodied in Toni Morrison's *Beloved*—Gordon shows how the visible and the "barely present" are intimately related. She writes, *"Visibility is a complex system of permission and prohibitions, of presence and absence, punctuated alternately by apparitions and hysterical blindness"* (1996:15; emphasis in original).

According to Gordon, ghost stories and their revelation allow us to more fully comprehend the complexities of social life. She argues that it is so axiomatic as to be banal that sociologists recognize domination and resistance as "basic and intertwined facts of modernity" (Gordon 1996:193). Yet this recognition masks the deeper implications of how power operates, as well as "the always unsettled relationship between what we see and what we know" (194). By pursuing a sociology of "not only the foot-prints but the water too," or that which is not always observable, Gordon offers a valuable tool for understanding "the spellbinding material relations of exchange between the defined and the inarticulate, the seen and the invisible, the known and the unknown" (200). She suggests that not only should we attempt to recognize ghosts, to bring them into the light, but we should also engage with them, speak to them, in order to find out what they can tell us about "the degraded present"—a present that is marked by their disappearance and our longing for them. To be haunted, then, is to contend seriously with the ghosts among us, with "the very tangled way people sense, intuit, and experience the complexities of modern power and personhood." (194).

Renowned feminist science studies scholar Donna Haraway, too, has significant things to say about visibility. In a discussion about "the statistics of freedom projects"—those projects aimed at directing knowledge toward social change and justice—Haraway interrogates the work of the "invisible fetus" in the politics of reproduction, which she asserts are "at the heart of questions about citizenship, liberty, family, and nation" (1997:189). She suggests that the fetuses and babies we do not see are as significant as the ones we do. Referring to Nancy Scheper-Hughes's (1992) eloquent analysis of infant mortality in a sugar-plantation region of the Brazilian Nordeste, Haraway raises provocative questions that have, in part, inspired our project: "In a world replete with images and representations, whom can we not see or grasp, and what are the consequences of such selective blindness. . . . How is visibility possible? For whom, by whom, and of whom? What remains invisible, to whom, and why?" (202). Like Gordon, Haraway urges us to focus our attention on the missing images in representations of reality and truth, considering these crucial to the quest for justice and freedom.

Sociologists who consider themselves symbolic interactionists, meaning that they focus on the ways people act toward the objects and situations that have meaning for them and on the interpretive and interactional bases of social organization, have also been interested in invisibility, specifically with the work that connects the visible to the invisible. Sociologist Susan Leigh Star, writing about Anselm Strauss's legacy, suggests, "the visible things are actions, stuff, bodies, machines, buildings. In social science, as everywhere, we are constantly wrestling with the properties of visible things: they are many, they are resistant to our attempts to change them, they clutter our landscape everywhere. In facing the tyranny of blind empiricism, however, we temper the clutter of the visible by creating invisibles: abstractions that will stand quietly, cleanly, and docilely for the noisome, messy actions and materials" (1991:265). She argues that the central insight of Strauss's sociological research is that the visible and the invisible are "dialectically inseparable." They are held together by the glue of social action in the form of work. Presaging Gordon's analysis, Star argues that "to do a sociology of the invisible means to take on the erasing process as the central human behavior of concern, and then to track that comparatively across domains. This is, in the end, a profoundly political process, since so many modern forms of social control rely on the erasure or silencing of various workers, on deleting their work from representations of their work" (281).

Computer scientist Bonnie Nardi and communications scholar Yrjö Engeström (1999) present very similar ideas in their introduction to a special journal issue focused on structures of invisible work. They address the valuable and somewhat vexing question of when the invisible needs to be made visible. As they argue, "Visibility and invisibility are neither good nor bad in themselves. There may be costs to revealing or concealing expertise and work. . . . Visibility and invisibility are not monolithic quantities; they are relative to various perspectives within an organization" (3). Star and Strauss echo this point in an article in which they suggest, "On the one hand, visibility can mean legitimacy, rescue from obscurity or other aspects of exploitation. On the other, visibility can create reification of work, opportunities for surveillance, or come to increase group communication and process burdens" (1999:10). In other words, the concern is not necessarily to increase visibility for the sake of greater clarity alone, without attention to the costs and consequences. Rather, it is to recognize, as Star and Strauss suggest, that "the relation between invisible and visible work is a complex matrix, with an ecology of its own. It is relational, that is, there is no absolute visibility, and illuminating one corner may throw another into darkness. For every gain in granularity of description, there may be increased risk of surveillance" (24).

And with surveillance often comes danger for those bodies caught up in silent, sometimes secretive aggregates. Historian Achille Mbembe, building on Foucault's notion of biopower, defines sovereignty as "the capacity to define who matters and who does not, who is *disposable* and who is not" (2003:27; emphasis in original). He suggests that late-modern colonial occupation combines "the disciplinary, the biopolitical, and the necropolitical," with states congealing around specific "terror formations"

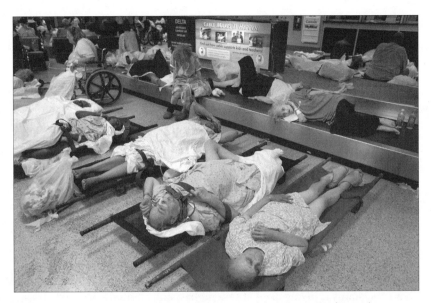

Bodies displaced by Hurricane Katrina. Photo Credit: Vincent Laforet/*The New York Times.* Source: Redux Pictures.

(27). Human beings become subjects not merely through self-care and biopolitics but also, and most consequentially, in the struggles through which they confront death. Mbembe's concern is not with sovereign powers whose project is autonomy but with "*the generalized instrumentalization of human existence and the material destruction of human bodies and populations*" (27; emphasis in original). Modernity and terror go hand in hand: sovereignty is the power to kill or let live, and he who has the biggest weapons—including the weapons of representation—rules.

Cultural studies scholar Henry A. Giroux (2006), writing about Hurricane Katrina, proposes the term "biopolitics of disposability" to describe necropolitical phenomena. Echoing theorists such as Italian philosopher Giorgio Agamben who contend that the regime of the detention camp is the definitive—and dangerous—biopolitical framework of our time, he writes, "The dialectics of life and death, visibility and invisibility, and privilege and lack in social existence that now constitute the biopolitics of modernity have to be understood in terms of their complexities, specificities, and diverse social formations" (181). Indeed, such a project is what we are doing in this book. Giroux suggests that images of the devastation brought by the hurricane peeled open the façade of the American dream, revealing the viscera of poverty, racism, and disadvantage at the core. In breaking through "the visual blackout" of social stratification, the Katrina images told us something about the hidden communities and corners into which the forgotten, abandoned, lost, and "wasted" humans have been shunted.[3]

As the ideas of Mbembe, Giroux, and others indicate, race, in particular, is crucial to an articulation of necropolitics, with racism historically functioning to regulate the distribution of death—and, we would add, disease. Discussing slavery as a historical instance of "biopolitical experimentation" and the origins of modern terror, Mbembe argues that slaves were kept alive "in a phantom-like world of horrors and intense cruelty and profanity" (2003:21). Life is subjugated to the power of death, marking necropower and necropolitics as necessary concepts to grasp changes in the contemporary world of terror formations. As Mbembe writes, "Weapons are deployed in the interest of maximum destruction of persons and the creation of *death-worlds,* new and unique forms of social existence in which vast populations are subjected to conditions of life conferring upon them the status of *living dead*" (40; emphasis in original).

These living dead, to borrow from Gordon, inhabit a kind of ghost world. And their presence haunts us.

Tracking the Ghosts: An Overview of the Book

Our theoretical project in this book, the recuperation of missing bodies and the circumstances of their erasure, presents a certain methodological quandary: How do we enable bodies that are muted in or by public discourse to speak in their own terms? In other words, how do we as scholars come to understand the missing if their lives and indeed their very corporeal essence are systematically ignored, erased, unseen, or missing in action? How do we "measure" the absent subject? And how do we "operationalize" invisibility? These are more than strictly methodological questions. These are political questions, ones that are deeply interconnected with our theoretical project of innovating an *ocular ethic*. What such an ocular ethic might enable us to do is to forge a new legacy of looking: one that refuses to assign political value to some bodies at the expense of others, one that treats "human subjects" in the fullness of their lived, embodied experiences.

The ethnographically grounded ocular ethic we propose is a strategy and perspective employed throughout this book. This ethic combines our feminist politics with our unique vantage points as sociologists of science, medicine, and technology. The dimensions of the ocular ethic include focusing, magnification, and visualizing. By focusing, we mean the process of drawing our attention and analytic gaze to the often-marginalized bodies, individuals, and groups in social life. While we are attentive to the biopolitical mechanisms that erase their voices and bodies, we also argue that children, dead babies, women, and people with diseases must be seen in situ and on their own terms. The act of seeing them, of focusing on them in a critical way, is an ethical responsibility. The ocular ethic uses techniques of magnification, including ethnography, to reveal, resituate, and recuperate. Ocular lenses are devices that can magnify the image formed by the "objective" lens, and this then can focus our vision on a particular aspect of images in a photograph or tableau. Just as the power of the microscope can be enhanced by turning a dial to increase the degree of magnification of an object, we practice an ocular ethic of magnifying bodies and body parts that are hidden or concealed, sometimes deliberately so.

We are not, of course, unaware that the 1980s and 1990s witnessed a proliferation of studies of representation in U.S. academia. Indeed, we were trained during these years in new theories and methods of the visible by Adele Clarke, Virginia Olesen, Donna Haraway, and other feminist scholars. The so-called postmodern turn in feminism and emergent politics of visibility led to fruitful reconsiderations of gender, race, embodiment, and power. Yet ultimately, the American cultural studies project, dominated by literary scholars and unmoored from its sociological and political roots in British cultural studies of the Birmingham School, fell out of academic vogue. Privileging vision as a dominant way of knowing was seen as hegemonic by postcolonial feminists such as Trinh Minh-Ha (1989), while many social scientists and materialist feminists argued that seeing did not necessarily lead to social change. For example, feminist theorist Nancy Fraser's (1989, 1997) articulations of redistribution in place of representation stem from this decisive historical moment. We assert here that, although our project attends to visibility and representational politics, our approach is more akin to British cultural studies with its emphasis on empirical grounding of data and obdurate material realities. Even more, we propose that a lack of empirical exactitude likely helped to doom earlier, more abstract theoretical studies of representation.

As sociologists schooled in traditions characterized by a rich history of *empirical* investigation, including symbolic interaction, medical sociology, and feminist studies, we are not offering a free-floating theory of bodies.

What we offer instead is a grounded, ethnographic, discursive analysis of particular social patterns and practices related to embodiment. In our view, bodies are not merely or only texts or performances but flesh and bone, histories and entanglements, suffering and illness, capabilities and desires, life and death—in short, bodies are material and not just materialized. As social objects (Mead 1934) imbued with a kind of *thingness* (Merleau-Ponty 1962), bodies are enacted in and through social relations while also retaining corporeal agendas of their own.

Addressing these complicated theoretical and methodological issues requires an innovative approach. We use multisited ethnography here to locate, collect, organize, and analyze data on embodiment (Rapp 1999, Marcus 1995). Note that we are not offering a collection of discrete ethnographies of each category of "the missing" in our project. Rather, we deploy this comparative method strategically to reveal social processes of bodily erasure and exposure. Data sources include children's books, documentaries, clinical and scientific studies, popular media, policy papers, government documents, autobiographies, scholarly literature, Internet sites, nonprofit materials, fieldwork, interviews, and statistics.

Our six case studies are organized into three thematic sections: Innocents (chapters 2 and 3), which takes up the assemblage of contested meanings that accrue to babies and children; Exposed (chapters 4 and 5), which interrogates the attributable risk of vulnerable populations in biological catastrophes; and Heroes (chapters 6 and 7), which explores the gendered relationships among heroism, capitalism, and nationalism. Taken together, these empirical case studies reveal interconnections among invisibility, vulnerability, risk, and power.

Part I, "Innocents," begins with chapter 2, a critique of how information about childhood sexuality is socially (re)produced. Sex education in the United States—that particular constellation of discourses, practices, and images—is designed to teach children about "the birds and the bees." As a society, we have gained insight into children's sexual identity and the eroticism of children's bodies largely through psychiatry, based primarily on now-outdated Freudian ideas. The imagined libidos of children, and perceived risk to children and adults (think Lolita here), have shaped this body of knowledge and policy responses to it. One unfortunate consequence of this attitude is that social scientists have tremendous difficulty gaining access to children as research subjects to investigate the emergence of sexual selves. We explore what sexuality education might look like if policymakers, parents, and teachers recognized that "sex ed" has a great deal to do with youthful bodies engaged in actual erotic activity, from masturbation to intercourse. Why is it, we ask, that we repressively terminate the inquiry by allegedly claiming it to be, as Judith Levine (2002) argues, "harmful to minors"? We argue here that erasing the erotic bodies of children and replacing them with psychiatric abstractions and moral pronouncements, in part, has led not only to a dearth of valuable data about children's erotic lives but also to underground traffic in visual images of children's bodies. Such traffic, consumed by "predators" and others, obviously does not foster children's safety and health.

Continuing our theme of tracking threats to innocents, we turn in chapter 3 to configurations of infant mortality. When this issue is discussed in the United States, it is typically framed in demographic terms in which the problem is about a specific kind of mortality or measure—in this case, "infant"—that often marks or stands in for another problem such as "race" or "class" or "health disparities." Using statistical measures, we report on the rates of infant mortality, or the incidence and prevalence of child death in the first year. But, it is disturbing to note, the demographic register is almost never about *actual dead babies* or the terrible grief and pain of child loss. Thus, public conversations about infant mortality are hollow, disembodied, and abstract and so of course fail to generate passion and social movement. What is at stake, we ask, when American discourse about reproductive health and the future of the species avoids the reality of dead babies by instead highlighting infant mortality statistics?

In part II, "Exposed," we direct our attention to issues of embodiment and vulnerability in the context of "human security" threats. We begin in chapter 4 with an analysis of the HIV/AIDS pandemic as a "biodisaster" of epic scale. Critically analyzing the creation of epidemiological knowledge, we show how quantification is used to represent individual episodes of suffering and disease. Numbers come to represent and to predict aggregate risks, and can be transported across social settings in which the actuarial becomes central to state formation and action. We chart the shift in U.S. policy about HIV/AIDS from an emphasis on public health to one focused on national security, particularly in the post-9/11 era. In this new framing, unlike a previous era in which people with AIDS were considered dangerous vectors of transmission, in the contemporary moment the object of geopolitical concern is the "failed state." The security interests of those in the developed world hinge on the stability, or lack thereof, of nations devastated by the HIV/AIDS pandemic. In place of public health infrastructure concerned with the sick and abject, we get an intensively capitalized biosecurity apparatus concerned with the "terrorist" and his or her disintegrating nation of origin.

Resembling the dramatic, overblown flair of FOX News, promoters of the U.S. security apparatus would have us believe that terrorist cells are devilishly plotting to infect the nation-state through deployment of biological and chemical warfare. Of course, exaggerating certain spectacular risks can detract attention from other stealthy and insidious risks that are not widely marketed. For example, our bodies are vulnerable to a consistent but seemingly ordinary, often invisible threat: environmental toxins. These are less-thrilling hazards than bioweapons, yet they may thoroughly devastate organic life. Every day, human bodies are bombarded with harmful synthetic chemicals that degrade our biological integrity. Historically, this toxicity was measured through sampling the environmental agents of absorption—soil, air, or water. More recently, measuring this toxicity can be accomplished through extracting body fluids from human bodies. Human biomonitoring, the scientific enterprise of using biological data to determine toxic load and forecast population exposure rates, is a 20th-century innovation, retooled for the 21st century, which relies on data mining of human bodies. As we examine in chapter 5, different meanings are attached to the use of body fluids, specifically breast milk and semen, which offer a map of gendered power relations.

In part III, "Heroes," we explore connections among the visible and the obscure, the triumphant and the vulnerable, and masculinity and femininity in relation to the nation-state. In particular, we ask what types of bodies become iconic of a certain kind of American national identity, and how is this produced? Using the figures of über-athlete Lance Armstrong and rescued American soldier Jessica Lynch, we discuss what it means to inhabit a gendered, heroic body in a post-9/11 world. At the same time, we ask what bodies are displaced by the relentless focus on these photogenic heroes.

Popular stories about POW Jessica Lynch represent a familiar and compelling narrative. In the "wag the dog" orchestration and coverage of her rescue, she was portrayed as a kind of fragile, feminine national treasure in need of liberation by tough (male) American troops. As a combatant in the Gulf War, by most definitions she is a genuine hero; she won the Purple Heart, a Bronze Star, and a Prisoner of War medal. Yet Lynch's story, narrated in chapter 6, is complicated by the fact that she is simultaneously cast as a highly visible hero *and* as the princess unable to rescue herself. Her heroism is deeply compromised by her femininity and further unsettled by her refusal before Congress to allow her image to be manipulated in the service of war. Understood within the context of "women in the military," we juxtapose Lynch's tale with the invisibility of women soldiers' lives and deaths. For example, we unravel the intricate mechanisms of power at play in the revelations that women soldiers in Iraq died due to dehydration because they would not drink enough liquids; they were frightened of being sexually assaulted by male soldiers while using latrines at night.

Unlike Lynch, who is portrayed as weak and ultimately traitorous for rejecting the military's fairy tale about her, champion cyclist Lance Armstrong is rewarded for embracing his own manufactured mythology. Armstrong represents a highly visible specimen of red-blooded American masculinity (and, allegedly, of technical enhancement through "doping"), despite—or perhaps because of—his successful battle against testicular cancer. His "balls" operate as a symbol of recovered manhood, just as his multiple victories in the Tour de France offered redemption and hope to an embattled American public reeling from 9/11. As we show in chapter 7, Armstrong is victory embodied, and his image is reproduced over and over again on magazine covers, merchandise, and the Internet as an example of red, white, and blue heroism. Author of his own narrative, he has catapulted his fame into fortune, as well as being a significant presence in the anti-cancer world. Through his foundation and the LiveStrong Campaign, Armstrong has become the (white) face of testicular cancer—potentially obscuring the mundane lives, daily struggles, and profound suffering of ordinary people with cancer.

In chapter 8, the conclusions, we further articulate our notion of the ocular ethic and offer a working theory of corporeal visibility and invisibility. Our analysis of each substantive area presented here provides a framework within which to consider recurring patterns of erasure and magnification of certain human bodies, revealing sediments of gender, race, class, sexuality, and other configurations of power as they operate in society and popular culture. We offer this book both as an invitation to consider these issues and as a provocation to join us in more ably seeing and theorizing human bodies, their varied and consequential social and political spaces, and the implications of embodiment in the 21st century.

We want to offer a final introductory remark. As scholars of the biopolitical, we are acutely aware that we do not exist outside of these practices but are perpetually caught up within them. For example, we both birthed our children in hospitals and have used pharmaceuticals for a variety of purposes. Indeed, our very scholarship could be used in ways that extend the scope of biopolitics. What if, for example, in the process of making visible certain bodies, new rationalized practices are developed to control and monitor these bodies? We aim here to foster intellectual and political engagement that does not stand outside of the biopolitical, for this is impossible, but rather carves out creative spaces for alternative and resistant discourses. Alongside critique, we seek to uncover new ways of knowing and living in the 21st century.

PART I

Innocents

A child is a beam of sunlight from the Infinite and Eternal, with possibilities of virtue and vice—but as yet unstained.

Lyman Abbott, *The Outlook* (1898)

Western, Judeo-Christian ideologies are deeply informed by notions of youthful innocence. Propped up by cultural practices, beliefs about innocence are projected onto babies and children and, with the advent of prenatal visualization technologies, onto fetuses as well—not without contention. In Western frameworks, an innocent body is one that is unmarked, not guilty, and not tainted by stigma; it is the embodiment of purity. Always and everywhere in the West, innocents are conceived of as fragile, naive, and vulnerable; they are in need of protection. Innocents are also weak, ignorant, and tragic: innocence is something that will necessarily, eventually, be if not lost then corrupted, taken, or stolen. Ironically, social forces work to sustain the possibility of purity for as long as possible. Through multiple discursive and material machinations, most adults contort the world to keep children in the dark (i.e., a belief in Santa Claus and the tooth fairy, a reluctance to discuss sex)—or at least they work to create an enduring, potent *illusion* of innocence.

Thus, innocence is both temporal and ephemeral. Only all too soon, it slips through our fingers, fleeting away to be mourned—not by the innocent but by the already fallen. Inevitably, innocence will be corrupted as the innocent is deflowered, betrayed, spoiled, conquered, or cannibalized. The vulnerability of innocents to exploitation requires prophylaxis. In the West, conscientious social monitoring, typically but not only by parents, is required to maintain purity and value. Routinely, adults lament the loss of childhood innocence and seek to prevent encroachment of "culture" onto children's virtue.

Many social institutions are dedicated to the production of "the innocent": the family, schools, religion, and the state. Each institution capitalizes on fertile opportunities to marshal collective action in the name of preserving innocence. During the 1820s to 1860s in the United States, the emergence of innocence as a category for some women and children, typically white and middle- to upper-class, was coterminous with the rise of competitive capitalism. These women and children became economically superfluous, engendering "the cult of true womanhood" (Gordon 2002).

Through an elaborate analysis of visual representations of children throughout centuries, art historian Anne Higonnet (1998) suggests that the notion of the innocent child was an 18th-century creation emerging in art, literature, and popular culture as life expectancy rates increased, commerce was transformed, and religious ideologies shifted. This idea of children as not born into sin grew throughout the 19th century, with Lewis Carroll's proclamation of childhood as golden innocence. Ideas about children as innocent creatures emerged alongside notions about some women—those who were white and were not performing material labor—whose "work" became taken up with the care of their precious offspring.

Often, quite dramatically, innocents have been a powerful, persuasive hook used to encourage certain types of human behavior. We are told: "Do it for the sake of the children." That "it" might be to stop smoking or taking drugs, to work in an unsatisfying but bill-paying job, to remain in an unhappy relationship, or to go to war for the sake of the next generation. Many of us feel compelled to act out of our deep, collectively shaped concern for these unspoiled creatures. We are quick to note, too, that there are tremendous marketing opportunities attached to ubiquitous parental concern. As a result, law enforcement and the media track sex offenders, corporations sponsor events in the names of (some) daughters and sons, conservative pundits excoriate women who leave their children defenseless as they "selfishly" skip off to work, social service agencies tell (some) women they are terrible mothers if their children are harmed, and, always, capital conducts a brisk trade in goods and services related to children's well-being. We (parents) must buy, or they (our children) will be in jeopardy from violence, disease, isolation, terrorism, boredom, neglect, or social censure.

We argue that fear of the loss of innocence (and the loss of innocents) motivates social action that further silences these supposedly docile subjects. In many ways, we retain a sense of primitiveness or even nostalgia: when we think about children, they are figured as pre-social creatures, pristine in their being and flesh, clean of all wickedness. Their innocence is inextricably linked to their newness—the freshness of a brand new body whose expiration date is fast approaching. Innocents—embryos, fetuses, infants, and children—are direct from the source. These bodies retain a natural essence or foundational mooring on which to base claims about the rest of us. They are simultaneously evidence of humanity's inherent goodness (its "god-given" potential) and reminders of how far we—both individually and as a species—have already fallen from grace.

The notion of innocence almost immediately conjures up the lily-white, blue-eyed Gerber™ baby, revealing layers of racialized history embedded in widely circulating and profitable cultural images. This well-fed, cherubic, naked, smiling infant is biologically and materially linked to maternal innocence. The innocent babe is intertwined with white woman's position on the pedestal of (white) man. She is the virtuous maternal representation of the untainted maiden of motherly goodness and genetic integrity. Innocence, then, is an already deeply racialized *and* gendered category. Women of color have never been allowed onto the pedestal built by white men, and nonwhite children are treated in a dichotomous position to (white) "innocents." Children of color—especially African American children—are represented with shocking frequency as soiled, pathological, dirty, guilty, lost causes, oversexed, and destined for imprisonment or an early death.

Here, we position innocents as particular types of subjects in order to critically interrogate relationships among embodiment, innocence, social control, and social justice. Innocents are a category of person representing race, age, and sex (but not typically sexuality), and particular bodies are routinely deployed to reassert existing stratification systems. For example, the Gerber™ baby signifies vitality, health, nationalism, and hope for the future, while other children such as the "youthful offender" are already marked as unlikely to survive, much less thrive. It is notable that contemporary anxiety about children's loss of innocence is intermingled with broader cultural anxiety about the fragile dominance of the U.S. empire and threats to the nation.

We suggest that innocent bodies may also be missing bodies, in part because the production of innocence requires silence. The maintenance of innocence (and innocents) mandates a silent subject, one for whom somebody else, typically a parent, must speak. On a continuum of visibility, innocent bodies are highly visible and integrated into our collective conscious, usually as a cultural ideal toward which to strive. And yet children do not always or even frequently have a voice—despite ubiquitous deployment of their bodies. They are not allowed to be speaking subjects narrating their own lives, and even when they do speak, they are rendered inaudible. Furthermore, we do not ask them to participate in discursive or self-reflexive inquiry because we construct them as already incapable of speaking for themselves. Infants and children do not have the necessary power to get their needs met in particular ways, although with age comes some degree of autonomy and self-control. Immediately after birth, they do not have language skills nor are they capable of rational action—although they may be self-interested. Exactly when children acquire the capacity to think and act for themselves remains contested and culturally variable.

Multiple social worlds collude to fabricate a political, historical, and cultural reality of innocents needing protection, and of innocence needing to be preserved, at all costs. In this framing, parenthood is above all else about protecting one's offspring. Yet despite this massive effort, at the same time the innocents are in effect left out of the configurations of human agency. If one is innocent, one is clearly not enabled to be a decision-maker, agent, property owner, or sexual agent. Innocents as culturally prefigured never get a voice—which is not to say that children themselves do not have agency. As a result, we constantly talk through and about the innocents. Our sociological mission in the following two chapters is to track how childhood innocence and the bodies of the innocents are deployed in biopolitical domains.

2

Seen but Not Heard

Consequences of Innocence Lost

> We're going to have a society of dangers, with, on the one side,
> those who are in danger, and on the other, those who are danger-
> ous. And sexuality will no longer be a kind of behavior hedged
> in by precise prohibitions, but a kind of roaming danger, a sort of
> omnipresent phantom, a phantom that will be played out between
> men and women, children and adults, and possibly between adults
> themselves.
> Michel Foucault, "Sexual Morality and the Law" (1988)[1]

For the past five years, in addition to the leather-bound keepsake baby books Lisa updates for each of her daughters, she has kept another notebook that contains "Grace and Georgia's Vagina Monologues."
The book begins with Grace at age 4 taking a piggyback ride and telling
Lisa not to worry about her falling off because she was "holding on tight"
with her vagina. A few months later, as Lisa and her daughters were in
the bathroom chatting while Grace "went potty," Grace told her mom that
sometimes she used toilet paper to clean out her vagina. Lisa responded
that one of the great things about having a vagina was that it was part of
the body that cleaned itself and so Grace didn't need to clean it. Giggling,
Lisa's daughter rolled her eyes and replied, "Oh Mommy, how can the vagina clean out itself? It doesn't even have hands." Her sister's vagina also
makes appearances in the notebook. As Lisa was bringing a cup of ice
water into Georgia's room one hot night, her youngest daughter, then age
3, pulled her hands up from under the covers and exclaimed, "Mommy,
you have to smell this. It smells *so* good, like lollipops and popsicles." She
placed her hand under Lisa's nose for her to appreciate the odor of her vagina. Lisa handed her the sippy cup, nodding in agreement. And months

later, Georgia, after discussing anatomy in her kindergarten classroom, barged into the bathroom and insisted that Lisa reveal her clitoris before she would eat breakfast. Lisa agreed and did a quick anatomy lesson using her own body as the model.

As we recount these entries from Lisa's "vagina" notebook, there is a certain nervousness we feel at sharing such intimate details. American culture treats the combination of sex and children as potentially explosive, and thus it is scary to openly talk about the topic. Indeed, in an October 2007 *New York Times* article titled "Shh . . . My Child Is Sleeping (in My Bed, Um, with Me)," science writer Tara Parker-Pope explores the dirty little secret of bed sharing or co-sleeping between parents and children. Parker-Pope cites research that suggests that in the West parents often do not reveal they are co-sleeping for fear of stigma or criticism of their parenting. Yet it feels unquestionably scarier to reveal one's own young daughters' erotic experiences and genital anatomies. There is, of course, the maternal transgression of exposing them and, more troubling, the fact of sharing our interpretation of their stories without their consent. Sharing our private conversations about children's ("my children's") private parts invites the potential for speculation, criticism, and accusation, or even possible intrusion by the state. Of course, we have been conditioned to want to proclaim the innocence of our daughters, their simplicity and joy at discovering themselves and our own innocence, our innocuous intentions to discover them. But this proclamation is complicated by the fact that we live in a culture of sexual contradictions—sexual prohibition for children and, at the same time, hypersexualization of adolescent bodies, particularly female bodies, through pop culture. Our girls (ages 4–9 at the time of this writing) witness bodies of "big girls": girls only a couple of years older than them, dressed in belly shirts and lowrider pants, their mouths enhanced with lip gloss and eyes with glitter that is marketed strategically to the "tweener" generation and consumed by all others.

As many can attest, speaking on behalf of children is a tethered responsibility of parents and guardians, most especially mothers. Due to their physical vulnerability and inability to communicate through verbal language, infants and young children are wholly dependent on adults in order to satisfy and express their needs. In every sphere, society relies on adults as proxies for children. Children and infants can't feed themselves, change their diapers, register for school, obtain their own shots, vote, or own property. Commonly in the United States, institutions—the courts, the educational system, the medical industrial complex—speak on behalf of children, sometimes in collaboration with parents, but frequently not. Often there is conflict between the institutional and the parental interests of the child. Jurisdictional battles are waged over the child's body between and among family members and other institutional actors, such as schools. Parents, particularly mothers, are suspect.[2]

This imperative that adults speak for them has profound consequences for the rights of children. Indeed, the United States is one of two countries (the other is Somalia) that has not completed the ratification process as a participant in the United Nations' *The Convention on the Rights of the Child* (CRC 1989). The United States has had difficulty ratifying the CRC because political and cultural conservatives do not wish to empower centralized public agencies with authority over the family (Smolin 2006). Furthermore, the CRC promotes certain policies that are in conflict with policies in many U.S. states, such as the prohibition on sentencing juveniles to life imprisonment with no opportunity for parole. Since 1989, this international treaty established and marshaled resources to incorporate the full range of human rights—civil, cultural, economic, political, and social—to people under 18. As stated in UNICEF materials publicizing the CRC, children everywhere must be guaranteed

> the right to survival; to develop to the fullest; to protection from harmful influences, abuse and exploitation; and to participate fully in family, cultural and social life. The four core principles of the Convention are non-discrimination; devotion to the best interests of the child; the right to life, survival and development; and respect for the views of the child. Every right spelled out in the Convention is inherent to the human dignity and harmonious development of every child. The Convention protects children's rights by setting standards in health care; education; and legal, civil and social services.[3]

Signing the CRC means that nations are accountable before the international community to develop and administer policies that protect and foster children's human rights. As a means of monitoring participation, governments that have ratified the CRC submit regular reports on the status of children's rights in their countries. The ideological underpinnings of the convention consider children as fully human: "Children are neither the property of their parents nor are they helpless objects of charity. They are human beings and are the subject of their own rights."[4]

With respect to children, then, there exists a set of tensions. On the one hand, there has been considerable global progress expanding the rights of children, although work remains to be done in the United States. (Indeed, fetuses fare better than children in terms of benefits and aid, as evidenced by the Unborn Victims of Violence Act and Medicaid coverage under the SCHIP provisions.) Yet on the other hand, despite this broadening of children's rights, because of their dependent status, age, cognitive development, and lack of resources, children are largely unable to speak for themselves—or to be *heard* speaking for themselves. Some person or institution is always speaking on behalf of children, whether it is a parent, a teacher, a physician, or the state. There is something interesting and provocative about this relationship: What does it tell us, politically and culturally, when others speak for children? What is at stake, and what is fostered? What are the terms through which an adult's right to speak for a child is secured?

Nowhere are these questions more intriguing, and perhaps more momentous, than in the realm of children's sexuality. When others speak for children, as we will show, it is almost always in pursuit of an agenda that is not set by children themselves. This includes our descriptions of our own children's sexual selves, both in this book and elsewhere (Moore 2007).[5] Within this agenda, children are typically preserved in an ideal type of "innocence" that may not be, and quite often is not, reflective of their reality. What is at stake, then, is adults' ability to define needs and to set agendas for an individual child, a group of children, or childhood more broadly—with potentially negative and even tragic consequences for "innocents." This ability of adults to speak for children and to interpret their stories and experiences is enabled by age of consent laws, restrictions on voting, and political participation—not actually listening to children, not recognizing children as emancipated—and, for our purposes here, viewing them as being in "protective" status by institutional review boards (IRBs).

The Untouchables: Researching Children's Sexuality

This chapter is conceived through our experiences both as sociologists of sex and gender and as mothers of four girls, two each. Therefore, asking where information about childhood sexuality comes from—and to what purposes it is put—is motivated by personal and political entanglements. As sociologists, we are not given easy institutional permission or access to study children using qualitative methodologies such as interviews and ethnography. As working mothers engaged in feminist politics and scholarship, our daughters' sexual bodies are potentially mined for sociological data: What relationship do our daughters have to their bodies? How do our daughters define their sexual selves? How do they narrate their understanding of their bodies? And more specifically, does the misogyny of the larger culture seep into our daughters' self-concepts and images? Yet also writing as mothers—middle-class, white, scholarly mothers—we have access in ways that dads may not or strangers are forbidden. Our own sexual innocence as seemingly safe mommies enables us to ponder our children's sexual innocence—but only to a point.

Similar to artist Sally Mann using the bodies of her naked children as (controversial) subjects of her photography, as parents and academics we interact with the many social worlds around us, and both intentionally and unintentionally our daughters inform our intellectual work. We must navigate a world that sexualizes children for marketing and consumption. To enable us to critically attend to this task, we seek balanced approaches to understanding the complexity of children's embodiment. We want both freedom and innocence for our children, yet also bodily autonomy and a sexual life that is rich, safe, and rewarding. What we see missing from the literature on childhood sexuality are the real, lived voices and experiences of children. And our limited access to collect such data from children because of the stigma attached to sexuality research and efforts to protect children means that the knowledge is limited and partial.

As methodologists M. Elizabeth Graue and Daniel J. Walsh astutely acknowledge in their text *Studying Children in Context: Theories, Methods and Ethics* (1998), qualitative and quantitative methods have long viewed children as *objects* of scientific inquiry instead of considering them as research *subjects*. In this type of research, children are seen and not heard. Children are placed under the microscope, poked and prodded rather than allowed to explain their thoughts and feelings. Furthermore, "little if any attention is paid to the contexts in which children live" (1). The ability of researchers to collaborate with children on meanings about their lives and their experiences is curbed by cultural anxiety and professional standards. And clearly, there are limitations on positioning children as subjects. Chronological age and cognitive development do matter in terms of a child understanding herself and developing the language capacity to evaluate and explain her experiences. Moreover, as adults speaking to and analyzing children, there is a translation problem. Children explain things and narrate their lives in terms and metaphors consistent with their own logic, explanations lost to most adults as we "grow up." Researchers must be constantly vigilant to balance the very uncertain tension between children's "reality" (or multiple, shifting realities) and adults' imaginings about that reality. Yet while we cannot treat children as if they are just adult research subjects but smaller, we want to suggest that it is dangerous to carve out, in the name of protection, a preserved territory accessible to only a few.

With respect to children and sexuality, scientists and scholars from almost all fields have conceptualized the child and his or her activities and experiences, as a particular, surprisingly narrow type of visible body. Due to Western hegemonic control over the human sciences, disciplinary knowledge about children's sexual bodies is filtered through a specific lens. Take cultural anthropology, for example. The variety of sexual behaviors and variability of sexual beliefs and realities held by children have been "sanitized" in their export and translation to the West. In his cross-cultural survey of scholarship on children's sexuality, José Nieto found that "the Western taboo against the sexual relationship of an adult with a child has been labeled as child sexual abuse. Generally, this taboo resulted in the omission or eradication of such behaviors from anthropologic ethnographies that documented the cross-cultural evidence of these activities" (2004:462). So such instances disappeared from the research, or became invisible. But just because they weren't being written about did not mean they weren't happening.

Cross-cultural research demonstrates how ideas and practices with respect to children's bodies and sexuality are not stable or monolithic. Rather, mainstream and normative ideas about children as sexual beings change, depending on cultural circumstances and beliefs. There is variability of children's sexuality that may be read differently in different contexts. But at least in the contemporary Western context, since gatekeepers are limiting who can do research with children and the research methods used, we are only able to interpret and understand partial aspects of children's embodied lives.

Experts overlook a great deal of data from actual children, such as previous anthropological research on children's experiences of adult/child sex rituals or, in the United States, children's experiences of pedophilia. Nonetheless, even though lived experiences of embodied children may be missing, children as a population continue to be deployed to prove the "truth" of human nature. As we pointed out earlier, children are thought

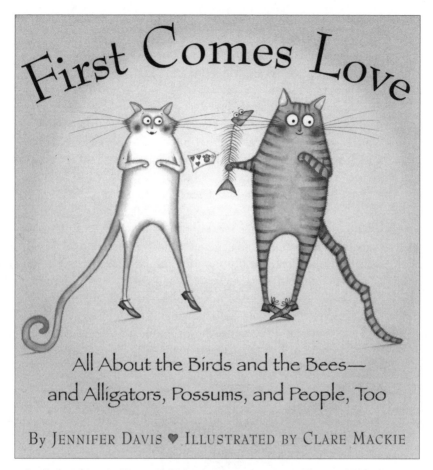

A lively example of a "facts of life" book for children. Jennifer Davis, *First Comes Love: All about the Birds and the Bees—and Alligators, Possums, and People, Too* (Workman), 2001.

to bear the essence of what makes us human. As film and media scholar Henry Jenkins has noted, "When we want to prove that something is so basic to human nature that it cannot be changed, we point to its presence in our children" (1998:15). Children are often constructed as some form of pre-socialized creature that exists as only a biological being and thus remains unmediated, animalistic, primitive, and naive. And we often lament that, as they mature, our children are "corrupted" by social forces. As Collette Granger states, "Significantly, the 'innocent' child is located in opposition to the adult. A deficiency of 'adult' knowledge, alongside a lack of awareness of the need for and ability to handle it, means adults must determine the need and bestow (or withhold) the knowledge. This opposition is particularly evident in the area of sex, since implicit in the innocence constructs are notions of childhood sexuality or sexual curiosity as inappropriate, even absent" (2007:7).

As social scientists, we wish to interrogate how (and why) children are constantly discursively produced as innocent—as we ourselves have done in this chapter—in the absence of actual information about their behaviors and experiences. Teasing apart these issues is the work of qualitative social science. Researching a previous book, Lisa had wanted to interview children about human sperm and their reactions to books written for the age group of 4–10 year olds (Moore 2007). She sketched out a study design involving focus groups after reading one of these books to an audience of children. But her informal inquiries to the Institutional Review Board (IRB) at her academic institution were met with such strong admonition against the possible "risks" to subjects and the "unethical" nature of the research design that she very quickly abandoned the notion of using children as subjects. So Lisa had no data about how children actually respond to books written for them (except for her own daughters). Instead, the grounded theory she constructed about children's sex education literature and human sperm was based in part on how adults imagine children read these books.

What are the grounds for "protecting" children? Although abuse of human research subjects has occurred throughout history, in the aftermath of World War II abuse of human subjects, specifically during "medical" research, was systematically addressed at the Nuremberg trials (Beauchamp and Childress 1989). The U.S.-administered military tribunal opened criminal proceedings against 23 leading German physicians and administrators. Nazi physician Josef Mengele's infamous experimentation, particularly on children's bodies, is often cited at the Nuremberg Doctor's Trial of 1946–47 (Lagnado and Dekel 1992). The creation of the Nuremberg Code of Ethics adopted in 1948 was the first modern legal attempt to establish ethical standards for modern bioscientific research, and the effects have been far-reaching. By 1953, the U.S. National Institutes of Health (NIH) required that researchers protect human subjects through the oversight of a review panel. The U.S. Public Health Service (PHS) followed in 1966 with regulations extending this review requirement to all "extramural" research supported by the agency. As the scope of government funding of bioscientific research grew, so did the creation of review boards for government-sponsored research—these institutional review boards (IRBs) were created in hundreds of institutions.

The creation of IRBs, however, did not mean that abuse did not occur. An especially egregious example was the PHS-run Tuskegee Syphilis Study of 1932–1972, in which 600 black men from Macon County, Alabama, were monitored and many were *not* treated for late-stage syphilis in order to collect data, even after treatment became available.[6] Outcry over the Tuskegee Study, as well as other examples of unethical conduct (see Beecher 1966), prompted Public Law 93-348, which established the National Commission for the Protection of Human Subjects of Biomedical and Behavioral Research. In 1979, the commission published recommendations known as the Belmont Report.[7] In order to conduct research at most universities, like our own, scholars must undergo training (web based or classroom) in the Belmont principles. Even though we as social scientists are not conducting *biomedical* research on human subjects, any interaction with human subjects generally falls under the IRB purview.

Human subjects are defined as "living individuals about whom an investigator obtains (1) data through intervention or interaction with the individual or (2) identifiable private information (45 CFR 46.102(f))." Some examples of social scientific research for which it is difficult or impossible to receive IRB approval are the following:

- A linguist seeking to study language development in a preliterate tribe was instructed by the IRB to have the subjects read and sign a consent form before the study could proceed.
- A political scientist who had bought a list of appropriate names for a survey of voting behavior was required by the IRB to get written informed consent from the subjects before mailing them the survey.
- A Caucasian Ph.D. student, seeking to study career expectations in relation to ethnicity, was told by the IRB that African American Ph.D. students could not be interviewed because it might be traumatic for them to be interviewed by the student.[8]

Scholars of bioethics have long argued that IRBs are often more concerned with protecting the institution rather than the subject. This act of protecting the institution is, we contend, part of a larger project of neoliberalism, an advancing and expanded legalism that chills some research—especially in the realm of sexuality—rather than protecting actual human beings. At the same time, this project bolsters profitable university-corporate relations, most obviously in the realm of pharmaceuticals, information technologies, and weapons. What we mean by "neoliberalism" is a shift in social relations that is characterized by government and economic institutions collaborating to create opportunities that favor growth in the service of large global and transnational enterprises. This shift has obvious and tacit effects on economic and social relations. As social theorist David Harvey (2007) has suggested, neoliberalism produces the doctrine of market exchange as an ethic in itself guiding human action. Protecting the fiscal or corporate interests of an institution becomes the underlying strategic operating principle that is disguised within the more publicly stated goal of protecting individuals.

With respect to bioethics, IRB provisions may slow biomedical research, for example, but they ultimately do not stop it; the provisions thus have no real "teeth" as anthropologist Barbara Koenig noted in Monica's study of fetal surgery (Casper 1998). It is remarkable to us that IRBs will approve experimentally cutting into a woman and her fetus or placing a monkey heart into a human but will not approve research asking a child about a "fact of life" book which is seen as "too risky." This chapter is a deliberate attempt to intervene in these debates about ethical parameters of research on children, calling into question who decides what may harm a child and revealing the deeper structural relations that pattern cultural ideas about children's sexuality.

The private sphere via the institution of the family interacts with social and economic institutions on behalf of children. As we suggest throughout this chapter, the configuration of children's bodies in each of these institutions relies on and fosters particular notions of innocence, harm, parental responsibility, and family itself. Families are constructed here as much as sexuality is, but these are fragmentary constructions and reveal power in many forms. Specifically, we aim for a critical analysis of the creation and deployment of children's sexual bodies as deeply interconnected to the policing of access to children and their parents, and to establishing the norms of engagement with children. For example, with respect to IRBs, parents are often asked to consent on behalf of their children if they are participating in research. Because of the presumed harm to children, based on what we read into and onto children's bodies and minds, institutions such as the IRB act as a gatekeeper and limit access to child informants. Indeed, the construction of a consenting research subject always presumes an autonomous thinking subject—the quintessential bounded rationality devoid of emotionality (e.g., Ramazanoglu and Holland 2002).

For example, regulations from the National Institutes of Health require valid informed consent be obtained by "disclosure of relevant information to prospective subjects about the research" and that subjects comprehend the information and provide "their voluntary agreement, free of coercion and undue influence, to research participation" (National Commission 1979).

The age status of children, however, complicates the ability of a research subject to assent or consent to procedures. The code of Federal Regulations, Title 45, Part 46, of the Department of Health and Human Services Policy for the Protection of Human Research Subjects defines children as "persons who have not attained the legal age for consent to treatments or procedures involved in the research, under the applicable law of the jurisdiction in which the research will be conducted." When a subject under 18 is capable of giving meaningful assent, IRB guidelines require that he or she, *in addition to* the parent or guardian, should be asked to sign the consent or assent form. If the minor is not able to comprehend the consent form, but is old enough to give assent (approximately age 7), he or she may be required to sign an assent form, and the parent or guardian must sign a consent form.[9] Before the consent forms are drafted, the risks to a research subject must be adequately parsed. Yet the notion of "risk" is subjective, particularly when the exposure—for our purposes, talking about sexual embodiment—is difficult to measure. This bioethical tenet to "protect" may in fact silence the voices of toddlers and children, as well as people with developmental disabilities or those suffering from trauma.

Activist and journalist Judith Levine's controversial, engaging book *Harmful to Minors: The Perils of Protecting Children from Sex* (2002) extensively examines the fears regarding corruption of children or their "innocence" and shows how this limits the production of knowledge about children's sexuality.[10] Through her research, including interviews with educators, policymakers, health professionals, and parents done between 1996–2000, she "investigates the policies and practices that affect children's and teens' quotidian sexual lives—censorship, psychology, sex education, family, criminal, and reproductive law, and the journalism and parenting advice that begs for 'solutions' while exciting more terror, like those trick birthday candles that reignite every time you blow them out"(xxi). As others have also observed (e.g., Brumberg 1998), children are used to sell products through their own nascent sexuality at the same time as they are trained to be consumers of sexualized items. Levine asks us to consider Barbie dolls, JonBenet Ramsey, and Britney Spears as prime examples.

The ways that we *talk about* (as opposed to display and consume) children's sexuality frame it as such that it cannot possibly be positive or socially desirable: "at best, youthful sex is a regrettable mistake; at worst it is a pathology, a tragedy or a crime. In the secular language of public health, engaging in sex is a 'risk behavior,' like binge drinking or anorexia. In religion it is a temptation and a sin" (Levine 2002:137). And only those from legitimate disciplines, such as clinical social workers and psychotherapists, criminal investigators, and educational psychologists, are given access to children but in highly circumscribed ways.

Significantly, children may have questions of their own about sexuality and their bodies. However, most researchers must ignore the children themselves and ask parents, teachers, or young adults (for retrospective accounts) in order to gather data about childhood sexuality. In 2002, using a close-ended survey, public health researchers at the University of Toledo interviewed elementary school teachers about their experiences with questions related to sexuality in their 5th- and 6th-grade classrooms (Price et al. 2003). Among the most commonly asked questions were "When do boys/girls start puberty?" "How can a person get HIV?" and "What is a penis/vagina/testes etc.?" Clearly, some 11 and 12 year olds are thinking about their sexual selves and their sexual anatomies and are asking the adults closest at hand. Although only one-third of these teachers received any formal training in sexuality education, a majority of them responded to a range of questions about sexuality in front of the class. Are 5 to 7 year olds also thinking about their bodies and sexuality? How could we possibly know?

In short, what has become apparent to many of us in attempting to do research about children and sex is a certain imagined construction of what children's sexuality should look like. A child's body is portrayed as a vessel of innocence, and this innocence in effect erases their actual bodies, desires, and experiences. Similar to author J. K. Rowling's invention of Harry Potter's invisibility cloak, this cultural veil of innocence prevents us from seeing the child hidden within. The construction permeates schools, families, advertising, fiction, the medical industrial complex, the university research system, and even the playground. Moreover, the moral panic about children's sexuality enabled by the cloak, in our view, actually precludes us from fully examining the messy complexities of children's embodiment. We argue that it is false and dangerous to substitute this version for reality because it ignores the potential for alternate positive *and* negative interpretations of children's lives. We ask: How might we understand children's sexual and erotic bodies in an ethical way? What is the child's construction of her own sensuality, or embodiment? Where in the world can we find this information?

Because certain disciplines (e.g., sociology) are not able to attend to *actual* children's bodies and to use their methods of inquiry to triangulate research subjects, the process of revealing the social complexity of children's sexuality suffers. We are prevented from establishing a fully robust theory of children's sexuality and bodily integrity—indeed, from even engaging in the kind of research that would allow development of such a theory. A first step would be to investigate barriers to the research process, including the following questions:

- Who has legitimate access to children as informants about their own sexual development and embodied experiences? What are the terms of legitimacy?
- Who can produce knowledge about the sexual development and experiences of children's bodies?
- Are children even allowed to be informants and, if so, under what circumstances?
- Who can children talk to?
- Why is collecting this information important? Whose purposes does it serve?

These are important questions because in considering them we can engage in a conversation about the ethics of research with children and broaden the scope of those who can participate in this debate. We return to these questions at the end of this chapter to propose guidelines for qualitative research with children.

In the absence of a dynamic research process, other discourses (e.g., popular, lay, or folk) spring up to define children's sexuality. Children are discursively constructed in particular ways that retain and intensify the essence of their purity—the cloak of innocence has considerable traction in these formulations. So, too, does the seductive appeal of revealing what is beneath the cloak. Information about children's bodies, sexuality, and threats to innocence emerges from locations other than the social sciences. Alongside the missing children in sexuality research—Judith Levine's work excepted—a partial representation of children's sexuality and the operations of the cloak of innocence may be gleaned from a critical investigation of social institutions. In what follows, we chart the ways in which children's sexuality is defined as normal or pathological by the medical industrial complex, including the public health education system, and by the media. We begin with the stronghold of the medical industrial complex over access to children's bodies and the definition of children's sexuality. These are highly consequential framings.

Hairy Palms: Biomedicalizing Children's Sexuality

In an article titled "Normative Sexual Behavior in Children," pediatric researchers showcased a questionnaire aimed at mothers of 2 to 12 year olds as a way to establish a sense of normal development (Friedrich et al. 1998). Children were excluded if there were any "obvious" indications of sexual abuse. Some 880 children, through their mothers, were included in the study. Findings reported a decline with age in overt sexual behavior: younger children exhibited more nudity and greater frequency of touching themselves. As children's exposure to cultural norms of sexuality increased and were linked with privacy, their sexual exhibitionism decreased. The article also referenced other studies: "Rutter has reported a variety of sexualized behaviors seen in younger children, including the following: erections in male infants, 'orgasmic-like responses' in boys as young as 5 months, thigh rubbing by female preschoolers, exhibitionism and voyeurism with other children and adults by male and female nursery school children, undressing or sexual exploration games in boys and girls by age of 4 years, and asking about sex by boys and girls by the age of 5 years" (456).

In classical biomedical fashion, the method of establishing the "normal" child's sexuality involves pathologizing the "abnormal" (Canguilhem 1977, Foucault 1978), and nowhere is there a greater historical example of this than masturbation. Derived from Latin words meaning hand and defilement, masturbation has a long history of association with cultural, religious, social, and medical taboos. According to Judith Levine, "the gradual pathologizing of normative children's sexuality, that is behavior that most kids do, has become increasingly surveilled and labeled pathological. This has consequences not just for the behavior deemed 'deviant' but also for *all* children's sexual behavior. Each time a new category of sexual deviance is identified—or, we might say, invented—the entire scale of so-called normal behavior is calibrated a few notches to the right (and to the Right). Professionals' and laypeople's idea of what is okay for children, teens, or families slides in a more conservative, more frightened,

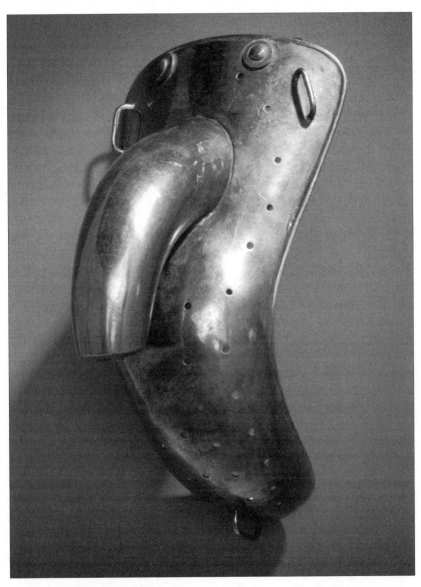

British male anti-masturbation device, 1871–1930. An example of the type of preventive device popular from the late 19th century through the early 20th century. According to D.E. Greydanus and B. Geller (1980), from 1856 to 1932, the U.S. patent office awarded 33 patents to inventors of such devices.

and more prohibitive direction, away from tolerance, humor and trust" (2002:48). Alexander Leung and William Lane Robson (1993) in an otherwise benign article from *Clinical Pediatrics* caution, "Masturbation in public is not appropriate and suggests either a lack of parental guidance or a significant behavioral disorder. . . . Repeated masturbation in public suggests the need for consultation with a child psychologist" (240). The nursing literature also provides instruction for clinicians in how to identify and manage functional and dysfunctional masturbation (Lidster and Horsburgh 1994).

In *Talk About Sex: The Battles over Sex Education in the United States,* sociologist Janice Irvine examines the social and cultural history of the Sex Information and Education Council of the United States (SIECUS). She explores why it is that since the late 1960s, social conservatives have been more culturally powerful than sex education advocates. Writing in 2004 as a participant observer in the local, state, and federal debates about sex education, Irvine finds:

> Since 2000, with conservative Republicans in control of both houses of Congress and the White House, advocates of abstinence-only sex education have advanced even more significantly on two related fronts: control over dissemination of sexuality information to the public; and control of federal funding for sexuality education and research. (Irvine 2004:xv)

Through deploying inflammatory terms and images, misleading information and depravity narratives as rhetorical strategies, social conservatives are able to steer public feelings and actions about sex education. Her book explores "how national advocacy organizations have scripted the public conversation on sex education through rhetorical frames which organize ambivalence, confusion and anxieties into tidy sound bites designed for mass mobilization."

The terrain for instructing children about sexuality has been landmined by social conservatives who work to impugn the integrity of individuals and organizations. It is no wonder that the recovery mission of acknowledging the existence of and then conducting research about children's sexual bodies is so anxiety producing. Irvine writes, "opponents of comprehensive programs targeted SIECUS, Planned Parenthood, and sex educators themselves, whom they described as the 'pornographers of the public school system.' In a speech during one local controversy, national right-wing activist Judith Riesman implied that sex educators tend to be pedophiles who enter the field to have access to young people" (123.)[11] (No, apparently, those are some priests.)

While research specifically about pre-adolescent children's sexuality is scant, there is a great deal of research about adolescent sexuality. Abstinence-only education somehow extends the fantasy of childhood innocence—as Levine observes: "If abstinence offers kids the freedom from growing up, it tends to give parents an equally impossible corollary, freedom from watching their kids grow up" (2002:108). Again and again, scholars and activists claim that researchers and politicians need to ground their claims in the real lived experiences of adolescents. For example, education scholar Jen Gilbert has suggested:

> Researchers have critiqued how youth, in relation to adults, are socially constructed as deficient, dangerous, in need of protection or caught in time (cf., Fine, 1992; Lesko, 2001; Pillow, 2004). In championing the interests of youth, this work has insisted upon young people's humanity, capacity for love and pleasure, and ability to take responsibility for their own and their partners' bodies. Indeed, these powerful critiques ask the radical question: what would it mean for adults to see adolescents as sexual subjects, and as having a right to experience the risks of sexuality, while also recognizing their responsibility to create the conditions for thoughtfulness, care and curiosity both in and out of schools? (2007:52)

Feminist social psychologist Michelle Fine's (1988) well-known scholarship on adolescent girls can be seen as a critical intervention to integrate a discourse on desire into sexuality research. In Fine's most recent collaborative research, she explores how even though data about adolescent female sexuality is readily available, the data is partial, not considering the lived experiences and embodied reflections of girls:

> Today we can "google" for information about the average young woman's age of "sexual debut," if she used a condom, got pregnant, the number of partners she had, if she aborted or gave birth, and what the baby weighed. However, we don't know if she enjoyed it, wanted it, or if she was violently coerced. Little has actually been heard from young women who desire pleasure, an education, freedom from violence, a future, intimacy, an abortion, safe and affordable childcare for their babies, or health care for their mothers. There is almost nothing heard from the young women who are most often tossed aside by state, family, church, and

school—those who are lesbian, gay, bisexual, queer, or questioning (LG-BTQQ), immigrant and undocumented youth, and young women with disabilities.(Fine and McClelland 2006:297)

This research cleaves well with our own project and further amplifies how particular bodies (and body parts) are missing from scientific data about social and biological events. The sexual anatomies of some women and girls may be deployed for strategic purposes by institutional actors, while at the same time, body parts such as the clitoris and processes such as orgasm are almost never mentioned (Moore and Clarke 2001). Rarely do most women or girls, especially those who are non-normative in their identities and practices, narrate their own embodied experiences.

Stop, You're Scaring Us: Children's Sexuality in Popular Culture

In an issue of a local New York magazine dedicated to children, a psychotherapist remarked, "What I hate about Park Slope, Brooklyn are the adults walking around with tense, frozen smiles while their eyes plead for reassurance that they have succeeded in capturing the American Dream. They're usually seen with a narcissistic, whiny kid in tow on one side and a high-strung Jack Russell on the other, while pushing Cleopatra in a thousand-dollar chariot" (Harris 2007:27). Parents, particularly middle- to upper-class parents, use their disposable income to foster reassurance of their "arrived" social status. Yet it is this very purchasing power that simultaneously fuels our collective anxiety; we attempt to purchase safety and instead end up buying (into) more fear.

Fear is marketed constantly to parents. Fear of dangerous invisible toxins in the home, fear of nonorganic food, fear of immigrants, fear of terrorists, fear of the Joneses getting ahead of you. One of the most dramatic sites of fear is early childhood education, the site of increasing surveillance over workers. Because of the vulnerability of the innocent child and the hysteria surrounding the "day care pedophile," childcare workers have self-reported that they consistently modify their labor practices to manage the culture of risk and anxiety (Jones 2003, Jones 2004). The possibility of sexual abuse of children is always and everywhere present in the intimate setting of a day care:

The spectral monster remains at the centerpiece in our safety policies, and fuels anxiety and suspicion about adult pleasure in children. While in the melodrama of the pedophile we like to position ourselves as the innocent and alarmed rescuers of vulnerable children, we are far more important than that to the scene. We invoke and reproduce the monster through positing children as vulnerable, by denying our somatic, emotional, and visceral pleasure in them, and by imagining the monster in every exciting touch and intimate act. (Jones 2003:247)

Spawned from the elevated sense of threat, websites proliferate to enable parents to track convicted sex offenders in their neighborhoods. Consider sites such as Family Watchdog (www.familywatchdog.us), a nonprofit venture started by a computer programmer in 2005, or the U.S. Department of Justice's Dru Sjodin National Sex Offender Public Registry (www.nsopr.gov), which enables a user to input her home address and immediately link to a Google map of registered sex offenders. Clickable color-coded boxes further stratify data to provide users with photos, home and work addresses, list of convictions, aliases, and descriptions of the offender. As a parent using this site, it is alarming how quickly one can get swept up in the culture of fear. The map of one's neighborhood is filtered through the map of colorful boxes of threats to your children.

And it is not merely "dirty old men," Internet predators, and day care workers who we are encouraged to be suspicious of around our children. In July 2007, the *New York Times Magazine* ran a story about youth sex offenders and the challenges of distinguishing sexual abuse from sexual curiosity. Due to the ways in which age of consent and sexual contact are defined, in many jurisdictions it is illegal for teenagers to have *consensual* sexual relations. For example, at the age of 17, Genarlow Wilson was convicted in Georgia for consensual oral sex with a 15-year-old girl. He spent three years in prison before the Georgia Supreme Court deemed his mandatory 10-year sentence cruel and unusual punishment. Several states have laws about tracking "sex offenders" under 14 years of age, but pending federal legislation may take this further:

In May (2006), Attorney General Alberto R. Gonzales proposed guidelines for the Adam Walsh Child Protection and Safety Act, named for a 6-year-old boy (and son of John Walsh, the host of TV's "America's Most Wanted") abducted from a Florida store and murdered in 1981. Among other things, the legislation, sponsored by Representative F. James Sensenbrenner Jr., a Wisconsin Republican, and signed into law by President Bush last year, creates a federal Internet registry that will allow law

enforcement and the public to more effectively track convicted sex offenders—including juveniles 14 and older who engage in genital, anal or
oral-genital contact with children younger than 12. Within the next two
years, states that have excluded adolescents from community-notification
laws may no longer be able to do so without losing federal money. . . .
Under the Adam Walsh Act, a 35-year-old who has a history of repeatedly
raping young girls will be eligible for the public registry, and so will a
14-year-old boy adjudicated as a sex offender for touching an 11-year-old
girl's vagina. According to the law, the teenager will remain on the national registry for life. (Jones 2007)

Fear is simultaneously quelled and aroused by the production of texts
for all ages about human sexuality. Sexual education books are one of
the ways cultural expectations are transmitted to children and relayed
to adults.[12] The niche market of "security moms" (white, middle-class to
affluent, suburban mothers) fuels the production and consumption of
books like *The Seduction of Children: Empowering Parents and Teachers to
Protect Children from Child Sexual Abuse* by Christiane Sanderson (2004);
Protecting and Parenting Sexually Abused Children by Rick Morris (2006);
*Children Who Don't Speak Out: About Children Being Abused in Child
Pornography* by Carl Goran Svedin, Kristina Back, and Radda Barnen
(1997); *Sexually Victimized Children* by David Finkelhor (1981); and *Protecting Your Children from Sexual Predators* by Leigh Baker (2002). We are
reminded here of anthropologist Gayle Rubin's (1984) prescient analysis of
moral panics surrounding sexuality, in which she argued that such panics
distract us from attention to structural inequality.

Our critique of the intensity of marketing fear to parents is consistent with the larger sociological analysis of the culture of fear. Sociologist
Barry Glassner, author of the 2000 book *The Culture of Fear: Why Americans Are Afraid of the Wrong Things*, demonstrates through meticulous
discourse analysis of the media how politicians and mass media produce
hysteria and panics without valid scientific evidence. These panics are often about statistically insignificant but dramatic events (such as contract-
ing Ebola or Bird Flu, or being killed in an incident of road rage) that
capture the Western public's attention while large-scale, enduring social
problems such as poverty are ignored. Ultimately, the threat of pedophiles
and the need to track one's neighborhood registered sex offenders become
imperative, while adequate, affordable, and safe day care or better schools
seem unrealistic.

How-to books are also abundant for parents broaching the subject of sexuality with their children. A sampling includes *Talking to Your Kids about Sex: How to Have a Lifetime of Age-Appropriate Conversations with Your Children about Healthy Sexuality* by Mark Laaser (1999), "Talking to Youth about Sexuality: A Parent's Guide" by Mike Aquilina (1995) and *Because We Love Them: Fostering a Christian Sexuality in Our Children* by Sheree Whitters Havlik (2005).

Havlik's book, in particular, offers a dramatic analogy of how sex education in an education environment is akin to putting a syringe in school age children's hands:

> Your child comes home from school and throws his backpack on the kitchen table. He opens it up and gets his books to begin doing his homework. As he pulls the books out, a syringe falls out. The syringe is in its original wrapper, unopened, nice and clean. Horrified, you ask your child where he got this syringe. "Oh," he says, "we had a presentation in health class today on not using illegal drugs. But they said that just in case we decided to do drugs anyway, we should protect ourselves from diseases and infections by using a clean needle. Then they passed out these needles." (9)

We have found no evidence that any school in the United States actually provides needles to children. But we do want to note the inflammatory rhetoric of using such an example in a sexual education book linking sex to drug use, marking both as clearly deviant behaviors.

The author also suggests an explanation about the perils of masturbation for parents to offer young children and teens:

> This is also why the Catholic Church teaches that masturbation is a sin. Masturbation does not give of oneself to another person. It is a self-gratifying act rather than a self-giving act. If sex is an expression of love, then masturbation says "I love me" rather than "I love you." It is a selfish act meant for self-gratification. Masturbation teaches instant gratification of sexual desires rather than a self-controlled appropriate expression with a spouse. (105)

By saying masturbation is bad and selfish, this text suggests that young adults forego sexual pleasure until marriage. Clearly, children and young adults must not make themselves feel good, but rather should be moral, and sexual pleasure must be given at a later date in the exclusive domain of heterosexual marriage. Undergirding this quote is the privileging of heterosexual prerogatives under the conditions of a marriage, in which sex is first an act of self-control and later a gift or duty to one's spouse.

In Lisa's previous research, whether in children's books, crime shows, or scientific literature, men, and by extension masculinity, are portrayed through representations of sperm cell behavior. Women, and by extension femininity, are portrayed through the actions of the egg cell. Eggs and sperm are given distinct personalities; they are the hero and heroine of these romance-like books. Just like in fairy tales, age-appropriate descriptions of how sperm jostle to be the best mirrors the idea of the singular prince in shining armor who will save the day. Further, she found that in children's books human reproduction is explained in conventional terms that reproduce idealized images of family, sexuality, and childhood (Moore 2007).

The ubiquitous presence of heterosexuality establishes that children's books are engaging in what sexuality scholars would call the cementing of "heteronormativity." Indeed, the child is mobilized in pursuit of heteronormativity and its collective future (Edelman 2004). By "heteronormative" we mean the processes by which heterosexual relations (that is, genitally born boys who become men having sex with genitally born girls who become women) are produced as natural and reinforced as transparent and unambiguous. Heteronomativity is a powerful idea that shapes human behavior in that the concept refines and reproduces itself using social and cultural ideas to convince individuals of the inherent naturalness and thus superiority of heterosexuality and heterosexual forms of intimacy. As literary critic Eve Sedgwick (1991) points out, adults often panic that children's sexuality will be radically different from what is considered normal—in other words, heterosexual and nondeviant.

In sum, building on the biomedical interpretations of children's sexuality as universally accurate and complete, popular culture is replete with conversations about normality and pathology. Narratives and parables of healthy, normal, and well-adjusted children raised by healthy parents in safe environments are juxtaposed by the unrelenting threats that eagerly await unsuspecting innocent children. Stories of sexual dangers framed around the bodies of children are continuously produced and popularized through the church, the school, and the state and brought into our homes at prime time via the television set.

Honey, Come See Who's on TV!
Criminal Justice and Pedophilia Entertainment

Segueing from popular culture into the criminal justice system's construction of child sexuality might seem a bit strange. However, we wish to critically examine the new phenomenon of what we call "pedophile entertainment," a hybrid of pop culture and criminology.[13] An argument similar to that of feminist cultural studies scholar Sarah Projansky about the "ubiquity of representations of rape" whereby "rape is thus naturalized in U.S. life, perhaps seemingly so natural that people are unaware of the frequency with which they encounter these representations" can be made about the growth of child predation as a form of entertainment (2001:2). This form of entertainment is constituted by television programs including *To Catch a Predator* (referred to here as *TCAP*), recent films such as *Little Children, L.I.E, Notes on a Scandal, Mysterious Skin,* and the documentaries *Deliver Us from Evil* and *Capturing the Friedmans.* Obviously, these entertainment vehicles suggest that we are a culture fascinated by children's sexuality: we both want to see children as sexual objects and want to have (imaging having) sex with children and think that they might want to have sex with adults. And yet, we are also a culture that wants to catch "the bastards" that challenge and destroy our children's innocence. The consumption of such programs undermines the claims about children's innocence, as we air and watch the shows, in prime time, with children themselves. We scare (and perhaps entice) them with constant warnings of "stranger danger"—a game Monica remembers playing often as a child in Chicago.

Ostensibly a critical and journalistic examination of the dangers and horrors of adult-child sex, these programs offer viewers an exciting opportunity to engage as spectators in the pleasures of horrified moral outrage, simultaneously getting the somatic buzz of treading through the forbidden.[14] A taboo that is apparently imminent and ubiquitous, as the host of *TCAP* warns: "Make no mistake, whether there are 5,000 or 500,000 on the Internet at any given time, they are out there and they pose a threat to your child. These men typically don't stand out in a crowd. They are cunning and patient. They often have respectable jobs. In our investigations we have caught doctors, a rabbi, teachers, a lawyer, musicians, an actor and all sorts of businessmen" (Hansen 2007:8). Since 2003, *TCAP* has set up hidden-camera investigations in a range of cities and suburbs including Bethpage, Long Island, New York; Ocean City, New Jersey; Mira Loma, California; Murphy, Texas; Greenville, Ohio; Fortson, Georgia; and

Herndon, Virginia. Our critical reading of this series explores the manufacture and maintenance of the system of innocence. In 44-minute weekly episodes, the host introduces victims, heroes, and villains, all of whom assemble around an imaginary 11–15-year-old body.

TCAP, as a spin-off of *Dateline NBC,* is teamed with Perverted Justice, an on-line watchdog organization that works to document and curb online predation. Specific episodes are advertised to the audience with NBC reporter Stone Phillips promising, "Some of it is *explicit.*" Each episode guarantees to involve one or more of the following: sex with children as young as 13, descriptions of oral or anal sex with minors, sex with animals, or mother-daughter sex. This is criminal justice as popular entertainment, with pornography masquerading as reality TV. The general "script" of a *TCAP* episode follows a predator from first contact to arrest. Adult volunteers at Perverted Justice acting as decoys craft online personas of teenagers to snare predators. A meeting is then scheduled, during which the underage decoy will be "home alone" at a house outfitted with hidden cameras and recording systems (Farhi 2006).[15]

We examine *TCAP* as a case study of a specific form of hypersexualized entertainment. Chris Hansen, an investigative journalist and host of *TCAP,* wrote in his book about the program, "What would become one of the most successful series ever aired on *Dateline NBC* had an inauspicious beginning in Feb 2004: I was stuck in traffic on the Throgs Neck Bridge, en route to a sting operation in a suburban home on Long Island where potential sexual predators were about to arrive. My producer, Lynn Keller, was frantic. If the predators got there before I did, it could sabotage the whole operation" (2007:11). It was clear from early on that the decision to air *To Catch a Predator* was highly interwoven with a quest for ratings and the ego of the journalist: "*Dateline NBC* averages about 8 million viewers a broadcast. Some weak stories can often be padded out with other elements and additional interviews to make them succeed. But once again, this was make or break time. If sexual predators didn't show up, I had no story. *Dateline NBC* would have sunk tens of thousands of dollars into a boondoggle with my name on it" (20). Hansen is almost rooting for the predators—after all, without them he has no show.

Since the show's start, *Dateline NBC* and Chris Hansen have interviewed 260 men involved in 10 investigations in seven states. Our reading of *TCAP* likens the program to sanctioned pornography where the money shot, that climatic moment of pleasurable release that is a staple of pornographic films, organizes the cadence of each program. Indeed, even

Cover of book. Chris Hansen, *To Catch a Predator: Protecting Your Kids from Online Enemies Already in Your Home* (Tantor Media), 2007.

those intimately connected with the project reveal their own physical reactions to producing the program: Hansen narrates the first investigation with excitement, describing how "my heart was almost beating out of my chest." Dennis "Frag" Kerr, the assistant director of operations and chief financial officer at Perverted Justice, explains how it was "exhilarating that it worked out and they showed up and we got to expose them on national TV" (Hansen 2007:57).

In this reality program, the "stars" are men (to date there has not been a woman predator featured, although the program claims to not actively discriminate) who make contact with underage children (in this case, 11–15-year-old boys and girls) with the intent to have some type of sexual exchange. The decoys establish a meeting time and place with the man, and the hidden cameras go to work. *TCAP* creates a set—a rented home in a suburban neighborhood techno-fitted for surveillance. As a means to build a history of each man who comes to the house, *TCAP* dramatically splices video with transcripts of the "actual" online conversations between the men and decoys. This textual foreplay is slowly doled out to the audience as a way to craft a denouement, Hansen's confrontation and humiliation of the man for coming to the house: "What did you think was going to happen here?" In one instance, Hansen introduces a "memorable predator" by stating how he proposed "a bizarre sex act involving a cat and cool whip." The predator then strips naked at the request of the off-camera decoy and is open to the camera with his genital area blurred out. Other fuzzy shots include webcam images of predators masturbating in undisclosed and unknown locations.

Local law enforcement is also enrolled in the program as officers wait outside the "set" to catch the man once he leaves the house. Constantly reminding viewers of the invaluable public service of *TCAP*, Hansen explains, "many law enforcement agencies do not have the resources or experience to run such a sting operation, and wouldn't be able to conduct such a sting operation without PJ and the national media attention brought by a *Dateline* investigation, which often shows that a town is serious about combating this crime" (2007:6).

Hansen claims the show has also had an impact on children themselves:

> I'll never forget one afternoon I spent with a group of about a dozen kids between the ages of eleven and thirteen. I interviewed them as part of our first "To Catch a Predator" investigation. All of them were smart kids

from good families. Some were children of NBC News employees. Basically I wanted to show the kids some of the video from our hidden camera house and get their reaction to the way potential predators could go from the chat room to the living room, sometimes in a matter of hours. We rolled the tape and the kids were glued to the screen. But the interesting thing was that some of them automatically assumed the men were actors and that the tape was demonstrating what could happen instead of what did happen. (2007:223)

Clearly, the hypermediation of today's youth creates a dissonance between what is actually happening and what is fictionalized or dramatized. Indeed, the fact that Hansen purports that his highly orchestrated, hidden-camera television show should be consumed as reality to these children is remarkable. We draw attention here to his presumed lack of informed consent from an IRB for these interviews; this is not research per se but commodification of children's stories for the sake of entertainment.

The show has obviously filtered into the American popular consciousness as Chris Hansen has appeared on *The Tonight Show with Jay Leno, The Oprah Winfrey Show,* and *The Daily Show with Jon Stewart,* as well as spoofs from the opening of the Emmy Awards with Conan O'Brien (given the screen name "conebone69" for the skit), *Saturday Night Live, Studio 60 on the Sunset Strip,* and multiple YouTube parodies. Hansen even plays himself in an episode of *30 Rock* where the lead character learns of her boyfriend's criminality while watching the show.

To Catch a Predator is, at least on the surface, about catching criminals. But it is firmly within the genre of reality television, so the experience of pedophilia has a game show–like quality—similar to the shows *Top Chef, The Amazing Race,* or *Survivor*; each show competes with the last in topping or outdoing the previous segment or episode. It is highly produced, as indicated by the following quote from Hansen about how the show evolved:

On the Ohio investigation, for the first time we agreed to pay Perverted Justice for its work during our investigation. After all, I was getting paid, my producer and crew were getting paid, and PJ wasn't going to work for free forever. It mounts a large operation and we needed its contributors to continue our investigations. They did, however, spark some controversy. After extensive discussion we decided to pay a consulting fee. We knew we would be criticized by some, and indeed we were. The sheriff's department also wanted to deputize Del and Frag. Because of a quirk in local law, authorities said deputizing them would allow prosecutors to charge our visitors with a more serious crime. It's something we weren't crazy about doing and it opened us up to criticism in journalism circles that we were too cozy with law enforcement. At the end of the day, it's something we felt we could live with for the limited purposes of Ohio. (2007:151–152)

It works so well because the show carries the label of "protecting" children, actually empowering the series to exploit the pedophile angle all the more. Clearly part of the thrill is the potential of getting caught, and the shows dramatize the moment of capture—in some ways, this is another take on the pornographic money shot.

There have also been some literary best sellers that serve similar kinds of voyeuristic pedophilia urges. *The Alienist*, about a serial killer of boy prostitutes is fairly graphic. *The Lovely Bones, The Child,* and *Little Children* each have plot lines about the sexual abuse and rape of children. It would, in our view, be even more groundbreaking to investigate if *TCAP* actually excites spectators and incites pedophiles. Janet Staiger (2005), expert in media reception studies, encourages scholars to use interdisciplinary approaches, including psychoanalysis and discourse analysis, to unravel how exposure to mass media representations affects audiences. Her research suggests that viewers may be profoundly affected, in sometimes negative ways, by exposure to images of all kinds.

What is gained and lost by framing pedophilia in this way? Grabbing ratings, selling advertising, and producing a reality television program on sexual abuse and children is not *really* about predators and child sexual abuse—rather, it is a simulacrum, not a social justice movement that creates lasting change that will benefit children. The show is structured in such a way that actually impedes social change: How much more outrageous can the sexual deviancy with children become? How much younger can the victim be? How much more detail can be dramatically revealed in the next very special episode of *To Catch a Predator?* How are children's bodies positioned as "characters" in this entertainment enterprise? As each show tops the previous spectacle, with decoys in place of real children (rightly so), innocence is both consumed and deployed for entertainment. In this melding of the pursuit of "justice" (defined narrowly in criminalistic terms) with pop culture, actual children who may be victims of abuse and assault are not really present—they are literally missing from the televised representation of pedophilia. The men are caught and the makers of *TCAP* profit. But do children?

To return to our framing notions, clearly programs like *TCAP* add to the ongoing struggle of maintaining a child's innocence. It adds dramatic weight to the program that, in addition to corrupting an individual child, *TCAP* illustrates the threat to the nation's children by evil, predatory forces. This media angle is exploited with sensational flair to a ratings bonanza. What in the media spectacle is being addressed? Some people get arrested, but *TCAP* perpetuates the script of what we have established earlier about innocence. This is one case study of how the missing bodies are rendered missing as part of a larger project of reinforcing their innocence. Children are neither seen nor heard, and as they do not get to speak for themselves, we hold their presumed gratitude and innocence alive in our imagination, drawing on it for sustenance. The mutually reinforcing parental fears and media escalation are not about addressing the fear but capitalizing on it, presenting predation as marketed entertainment. As consumers, we are refueled to fight the forces of evil for another day.

Reclaiming Subjects

As sociologists, we want to talk about children's sexuality, and it is clear that journalists and pedophiles do, too—but not for the same reasons. This is not to say that we, unlike other sex researchers or reality TV celebrities, are immune to the messiness of exploring children's sexual identities and bodies. Just as *TCAP* is a more-complicated production and consumption junction than it claims to be, our proposition to engage with children as research subjects is also highly complex. It is erroneous to presume there is a clear border between entertainment and socially responsible research. We realize that research on children's sexuality could, in its end products, be used for purposes of pleasure. As some scholars have discussed (Plummer 1995), sexuality research may be pleasurable, and there might be somatic reactions to talking to children or adults about their bodies and feelings.

It is clear from our interpretation of certain social worlds that children's bodies are hypervisible in some venues at the same time that they are invisible and nonvocal in others. How is it that there is an abundance of representations of children's bodies in the world of child porn and pedophilia at the same time that there is a near absence in sociological research? What patterns of visibility and invisibility are at play here?

We are concerned with the structural barriers that disable full and responsible scholarship about children's sexuality. Obviously as a culture, we are curious about children's sexuality, but instead of figuring out an ethical means of investigation, what emerges in its place are psychological and medical studies detailing pathology, a criminology and "reality" entertainment nexus, and vigilante groups organized to "protect" (some) children. While the research enterprise is flawed in certain ways and the IRB process both insufficient and overkill at the same time, the entertainment industry is bottom line about profits. While sponsors and viewers can occasionally keep the networks in check—recall Janet Jackson's wardrobe malfunction at the Super Bowl—ultimately the dollar is king.

Often, we have debated each other and our colleagues whether it is acceptable to conduct any research about children's sexuality when it is obviously not just for children's sake. The research may benefit adult sexuality, expand the literature about children, and have merit for some vague broader "societal benefit" that has yet to be adequately delineated. Of course, children, like other human and nonhuman research subjects, may be harmed by the research process—as many IRBs claim, deploying the claim as a primary reason *not* to approve the research. But perhaps we need to interrogate the very terms of risk and harm, and who defines these. Certainly other interventions on behalf of children, even on behalf of fetuses and neonates, are approved for the advancement of research and human knowledge, despite extraordinary complications and often unclear benefits.

At present, in most contexts, children do not speak for themselves, because everyone else is busily making careers speaking *for them* while simultaneously pathologizing their behavior to normalize adult sexuality. And we must concede that we, too, want to get data from children for our own scholarly purposes. Yet we are limited because we are adults. It is our hope that, given adequate access, we might be able to participate in new discursive practices to enable children to speak. Clearly, we are not the first scholars to voice concern about the absence of children in research paradigms. In 1998, anthropologists Nancy Scheper-Hughes and Carolyn Sargent wrote:

> Children's voices are conspicuously absent in most ethnographic writing, where young people seem to behave like good little Victorians, neither seen nor heard. . . . If children's words, intentions, and motivations are missing in our ethnographies, their physical bodies are also absent, except

as sites of physical discipline, (genital) initiation, or sexual molestation. Despite a decade of anthropological research and writing on the body, almost nothing has been written on the body of the child, on children's body practices, tactics, and meanings. . . . Finally, while children have been—indeed, sometimes relentlessly—tracked, observed, measured, and tested, rarely are they active participants in anthropological research, setting agendas, establishing boundaries, negotiating what may be said about them. (13–15)

We add our voices to this call for change of the research paradigm. We also argue that there are consequences to the continued denial of careful inclusion of actual children into the design, implementation, and validation of social scientific research. It is our claim here that not attending to developing new methods for research about and with children hinders full participation in knowledge produced by children and about children's sexuality. These barriers to scholarship can lead to a proliferation of other forms, generally driven by market forces and multiple sexual desires, of accessing information and knowledge of children's sexuality—which may have very little to do with actual children and their needs.

For many political and legal reasons, as we argued earlier, someone always has to speak for children in institutional settings. Yet even though children cannot necessarily speak for themselves formally, we are arguing that ethical qualitative researchers may be best positioned to capture children's voices, experiences, and lives. At the moment, this is merely a theoretical project in the sense that we have not done this type of research here. Although we have tracked certain patterns about children's risk and vulnerability in discursive productions of innocence, this is not an empirical study of children's sexuality. However, an excellent outcome of this chapter would be to inspire researchers to conduct ethnographies of the missing, including studies of and with children. At the same time that we are critiquing barriers to research with children, we also want to issue a challenge (to ourselves and others) to seek new ways to listen to, represent, and share children's stories.

We feel that we must be cautious, though, in advocating for a research program about children's sexuality. Although we are making a broader argument about children and innocence, we cannot group all children together—a mature 9 year old may be able to comment on her body in ways that certain 14 year olds cannot. There are obviously power dynamics at play, along with cognitive, developmental and physiological differences, and linguistic and somatic skill sets that vary widely among age groups and maturity level. Our proposal for research guidelines for qualitative research with children about sex includes a commitment to age-appropriate research questions that enable and empower children to speak *for themselves*. Although conventions about reflexivity and allowing subjects to speak for themselves do exist in qualitative research, particularly in feminist methodologies, children are a different population from adults; this is not the same thing as saying they are "innocent."

In sum, as social scientists we are entangled in the complexities of modernist projects of knowledge production. At the same time that we are enabled by our feminist qualitative training in critical methodologies, we are also imprisoned by the positivist projects of revealing or discovering better or more accurate Truths. We believe that we should participate in the recovery operation of children's bodies and conceptualization of themselves in the realm of the erotic and sexual. However, due to structural limitations of how research is legitimized and approved in institutional settings, we are denied access. The register of protection, a kind of masculinist or paternalistic discourse framed by IRBs and funders, becomes the device used to guide all discussions about research with human subjects. But why is the register of the IRB about protection rather than enablement or empowerment? We contend that the protection discourse itself is what often shuts down the research or guides it in such a way that social change is obstructed. Modernist tools of research, these very methodologies in question, figure prominently in our next chapter on infant mortality, where we show how the methods of demography effectively conceal the social structures that undergird infant mortality as a conceptual category, thus erasing the actual bodies of dead babies.

3

Calculated Losses

Taking the Measure of Infant Mortality

Quantification is a social technology.

Theodore M. Porter, *Trust in Numbers* (1995)

In 2005, Americans were enthralled by a well-publicized, dramatic account of familial love, gender relations, extreme adventure, and child survival in a thoroughly inhospitable environment. No, this was not another prime-time episode of Mark Burnett's reality TV series, *Survivor*. Rather, it was a beautiful, forceful nature documentary, *March of the Penguins* (Warner Bros., 2005).

As portrayed by director Luc Jacquet, the film provoked national interest in the perils and promises of procreation. Safely tucked into the local cineplex with popcorn and soda in hand, human viewers were transported to the icy, dangerous terrain of Antarctica and the poignant travails of the Emperor penguins that live—and die—there. As one critic raved, "Given the advances in high-definition photography . . . [it] is exquisite to behold, but what makes this a truly great film is its emotional impact, as it tells a heartwarming love story from the coldest place on Earth" (J. Williams 2005).

By any measure, the journey of the Emperor penguins is indeed extraordinary. Traversing mile after glacial mile, "driven by the overpowering urge to reproduce, to assure the survival of the species," the penguins rely on instinct and "the otherworldly radiance of the Southern Cross" to locate their traditional breeding ground.[1] There, they pair off into monogamous couples to mate. The females of the species remain at the breeding ground just long enough to lay one solitary, precious egg. Then, leaving the eggs in the care of their male partners, the exhausted, hungry females embark on a perilous journey back to the sea where they will feast on fish and regain their strength—if they are not first eaten by predatory leopard seals. The male penguins, left behind with the fragile eggs, must ensure their own and their offspring's survival by any means possible. They cradle the eggs on top of their webbed feet, taking great care not to drop them, while they huddle together for warmth in a great, writhing mass of black, white, and orange. For many weeks, the males withstand the subzero climate, wintry blizzards, and their own hunger and fatigue in order to protect their young. At last, after two arduous months, the baby penguins hatch in a flurry of downy fuzz and high-pitched squeaks. But the fathers cannot feed the newborns, and some young will die before their mothers return from sea. When the females arrive, within days as in a carefully choreographed play, they bring food for the chicks, and the life cycle continues afresh. But the males must now make their own return journey from breeding ground to fish-filled sea, or they, too, may die.

As spectacle of the "natural" world, the film succeeds admirably. The cinematography is gorgeous; reportedly so grueling that the production notes, framed as an adventure tale all on its own, are highlighted both in the DVD and in the companion book (Jacquet et al. 2006). Yet the film also succeeds as an emotional journey filled with ups and downs, life and death, love and loss (penguins really do seem to grieve), and a gripping story arc narrated by Morgan Freeman.

A parable about contemporary family life with recognizable characters, the film relies on a conservative trope of heterosexual, monogamous love and the miracle of new life. That the male of the species is so involved makes for a provocative and compelling twist. Judging by the incredibly favorable reception of the film by critics and audiences alike, the anthropomorphism—that is, the human personification of the animals—of the penguin mommies and daddies is highly effective. In the words of one reviewer, "when it comes to understanding why the emperor penguins would go to such great lengths to mate and have babies, the inexplicability of human love may be the only comparison we have" (Zacharek 2005).

So much more can be said about *March of the Penguins:* as popular culture, from a feminist perspective, in terms of environmentalism, as documentary, and as children's fare. We want to focus specifically on the ways in which this film—essentially a superbly crafted tale of infant survival and death—captured the public's imagination. Television commercials featured families exiting the theater, commenting on how "awesome" and "wonderful" the film was. Parents described "laughing and crying" and professed delight that they had taken their children to see the film. Commentators described Americans' love affair with penguins: "Forget polar bears. This winter's 'it' critter is unquestionably the penguin" (Associated Press 2006). And while not an exact gauge, the popular Internet site www.rottentomatoes.com featured 155 fan reviews, with an average rating of 94%.[2]

And so we pose this question: In the throes of this penguin spectacle, in which average Americans and film critics alike were enchanted with what mommy and daddy birds must endure to keep their chicks alive, why was nobody visualizing *human* infant survival and death? Where was the movie about dead human babies with a swelling sound track and poignant images? Spellbound by a saga of risk and hope, life and death, and parental sacrifice on the icy killing fields of Antarctica—framed, no less, as a love story—were American viewers simply unaware that some 28,000 human babies younger than one die each year in the United States, many of them from preventable causes? Why are vulnerable baby penguins so visible in popular culture, while dead baby humans appear infrequently on our collective radar? In this chapter, we explore the biopolitics of human infant mortality, in which disembodied discourses of demography frame the problem as one of statistical risk rather than human loss.

Locating Infant Death in the United States

This analysis is part of a larger research project conducted by Monica: a genealogy of infant mortality in the United States that traces the complicated biopolitics of neonatal and infant death, including 20th-century classification, measurement, and preventive practices. The infant mortality rate (IMR) has been defined since the late 19th century as the number of babies out of every thousand born that die before the age of one, or the number of infant deaths per 1,000 live births. In 1906, British physician George Newman described infant mortality as a major social problem, defining "the death of infants in relation to a community." More recently, the World Health Organization (WHO) affirmed, "A healthy start in life is important to every newborn baby. The first 28 days, the neonatal period, is critical. It is during this time that fundamental health and feeding practices are established. It is also during this time that the child is at highest risk for death."[3] The IMR is one of the most frequent measures of the health of a nation and the failure of states to care for their citizens (Armstrong 1986).

Infant mortality is largely, but not only, a problem of developing nations. As any medical sociologist will tell you, a world map indicating
regions of poverty can be overlaid almost exactly with a map indicating
sickness. Health and disease are socially produced, and global inequality
is one of the leading factors in worldwide health disparities. This means
that the countries with the least amount of resources and the poorest people tend to have the highest IMRs. The further one travels down the list to
higher mortality rates, the further one moves from "developed" countries
in North America and Europe to those in the so-called Third World, or
the global south. In part, this is because maternal mortality rates are high
and women's health in these regions is poor. It may be true that sick and
vulnerable mothers make sick and vulnerable babies, but these connections have not often been measured or discussed by policymakers.

According to the CIA World Factbook, as of January 2007 the five
worst nations for infant mortality were Angola, Afghanistan, Sierra Le-
one, Liberia, and Niger, with bottom-ranked Angola reporting 184.4
deaths per 1,000 live births. At the top of the list are nations with the
lowest IMRs: Singapore in the number one spot with a rate of 2.3 deaths
per 1,000 live births, followed by Sweden, Japan, Hong Kong, and Iceland.
These numbers are clearly linked to structural resources.[4] For example,
in 2003, some 70% of the Angolan population lived below the poverty
line, while the number living below the poverty line in Singapore was too
small to register statistically. We want to note here, too, that the IMR in
Iraq grew 150% between 1990 and 2007 (Buncombe 2007). According to
the advocacy group Save the Children, 37% of the increase occurred after
the U.S. invasion of Iraq in 2003. Factors in rising infant mortality rates
there include electricity shortages, lack of clean water, deteriorating health
services, and soaring inflation.

Clearly, these statistics reflect deep and persistent global disparities. Yet
there is one surprise on every IMR list: namely, how badly the United
States fares in the rankings. The wealthiest and most powerful nation
in the world had, in 2007, a rate of 6.37 deaths per 1,000 live births—an
IMR almost three times as high as that of number one Singapore and just
slightly better than Croatia, Belarus, Guam, and Lithuania. (In 2006, the
United States' gross domestic product or GDP was $13.06 trillion.) The
IMR in the United States, as elsewhere, fluctuates to some degree annually. In 2002, the overall national IMR increased for the first time since
1958, to 7.0 deaths per 1,000 live births. There were 27,970 infant deaths in
the United States in 2002, and a majority of these were due to preventable

factors. Media coverage in the form of front-page exposés and features began to attend to this unwelcome rise in the IMR, focusing especially on the steep climb in Southern states.

Overall, the reduction of infant mortality in the United States was one of the major public health success stories of the 20th century and a hallmark of biopolitics in action. In 1900 in some U.S. cities, 30% of infants died before reaching their first birthday ("Achievements in Public Health" 1999). From 1915 through 1997, infant mortality declined more than 90%. Improvements in standard of living and nutrition, advances in clinical medicine, increased educational levels of parents, and better surveillance and monitoring of disease contributed to this achievement, as did a determined policy focus on maternal and child health as embodied in the Sheppard-Towner Act of 1921.[5] Also known as the Maternity and Infancy Bill, the act's goal was a reduction in infant mortality through provision of matching grants to states, inspection of maternity homes, and creation of facilities and programs. As the national IMR decreased in the wake of these structural improvements, so did public attention to the issue and government resources.

Yet, despite a dramatic improvement during the century, significant disparities in infant mortality have persisted across racial and ethnic divides. The history of infant mortality in the United States is, to a large degree, a history of immigration, race, poverty, and other structural inequalities. Many nonwhite infants, especially those born to African Americans, die at much higher rates in their first year of life than do white babies.[6] In 2002, the first year in almost a century during which the IMR in the United States increased, the mortality rate for African American infants was 13.9 deaths per 1,000 live births, a rate comparable with that of many developing nations. The percentage of newborns at low birth weight—a leading indicator of infant mortality—has risen steadily since 1984 and in 2007 was at the highest level recorded in three decades. African American babies are twice as likely to be of low birth weight than white babies and four times as likely to die from prematurity as white babies, and they are also twice as likely to succumb to SIDS, or Sudden Infant Death Syndrome.[7]

Regional racial and socioeconomic gaps go a long way toward explaining the low—some might say abysmal—U.S. ranking on the IMR charts. Further unpacking racial and ethnic categories, we learn from the CDC and other agencies that IMRs are also affected by cultural factors, unequal care, and unintended pregnancy, as well as the age of the mother, cigarette smoking and alcohol consumption by mothers, obesity of mothers, and education level of mothers. Importantly, fathers are *rarely* measured as contributing to infant mortality, further ideologically meshing mothers with their biological children. Even a cursory overview of these factors reveals complicated socioeconomic circumstances in which women and their babies struggle to live, which may lead to preterm births, neonatal illnesses, birth defects, and infant death.

What should be clear from this brief overview of infant mortality in the United States is the reliance on a language of quantification: that is, numbers. Infant death is articulated here in terms of rates, patterns, and risk; the infant mortality rate even has its own acronym: IMR. In narrating the scope of the problem, policymakers and health care providers rely primarily on a demographic or statistical register. This is precisely the kind of abstract discourse of which we are deeply critical in this book. Note that our discussion of the problem of infant death has not yet featured the bodies of individual babies who have perished or individual women and their families suffering a devastating loss. Narratives of grief and pain are thus far absent from our dry, technical summary—strategically so, because we want readers to pay attention. In the following critique of demography, the science most often deployed to make sense of infant deaths, we show that the IMR does not just *represent* health disparities. It is, in fact, fundamentally *shaped by* racialized disparities of income, status, and access to preventive health care.

Mortality in the Disembodied Aggregate

As feminist scholars, we are interested in tracking the measurable relationships between national wealth and infant mortality, between development and death, between racial formations and health disparities, and between geography and resources. Unlike the penguins of Antarctica, which are (framed as) relatively homogenous as a species, humans are widely variable, with vast differences along many parameters, including mortality. These measurable differences are collated into schemes and registers, which are then used to shape policy. Where the struggle of the penguins seemed to produce empathy and perhaps increased human concern for their habitat and conditions, aggregated data about human infant death appears to do the opposite: namely, drive rationalized policy devoid of empathy and care. Of course, a film shot in high definition will be more compelling than a pie chart or table. But we seek to understand why some phenomena are represented in an affective register, while other concerns are rationalized via abstract and numeric forms of representation.

As we introduced earlier, Foucault's *biopower* refers to regimes of knowledge and information about populations, or subjects in aggregate. In contrast to disciplinary power, which focuses on individual bodies, biopower is directed at the species. It requires the use of various techniques of knowledge collection and production—the human sciences—and may lead to new forms of governmentality and state power. Knowledge-production techniques in the human sciences target fertility, birth, life, disease, migration, sexuality, and death in order to foster and capitalize on the health and productivity of the social body. Crucial to the operations of biopower is normalization, or the regularized and standardized average against which all else is measured. Statistical measures rely on a technical norm and on deviations from that norm. Aggregate rates are useful for predicting the distribution of resources, as well as anticipating the flows and directionality of power. For example, when governments know the size of the employed population, life expectancy, and mortality rates, they can adjust the age of social security benefit distribution.

Demography (from its origin, "people writing") is the statistical study of human populations, and it has been integral to formations of biopower and thus to expanding the reach of the state. Using sophisticated quantitative methodologies and mathematical models, demographers offer systemic analyses of populations, including their characteristics, movements, changes, and distribution. The unit of analysis is the aggregate, not the individual. Topics of investigation include birth, death, disease, migration, density, growth, marriage, and aging, as well as the social processes that affect these. While demography is sometimes considered a subfield of sociology and many demographers work within sociology departments, it is more accurate to recognize the field as a distinct science with its own methodological standards and conventions of practice. Demographers have professional societies and journals (e.g., *Demography* is the official journal of the Population Association of America) and increasingly their own academic units. To cite one institution, the University of California at Berkeley has departments of both sociology and demography.

Demography is a thoroughly modern(ist) project in the sense that it reflects the Enlightenment principle that there is an objective truth to be determined through investigation. Practicing a positivist science, demographers rely heavily on their assumption that truth is quantifiable. To return to Berkeley, a sample of courses offered in the department of demography illustrates this bent toward the quantitative: economic demography; demographic methods: rates and structures; population models; computer applications for demographic analysis; mathematical demography; and so on. Moreover, the mission of the Population Association of America is "to promote the improvement, advancement, and progress of the human race by means of research with respect to problems connected with human population." In other words, in the association's view, macro data produced through these methods should be harnessed toward human achievement and uplift, not merely measuring populations but improving them at the same time. In short, demography is a model of the kind of rational calculation described by Foucault, central to processes of governmentality, including recordkeeping and classification.

From our perspective, one of the most troubling characteristics of demography is that its practitioners claim to be objective and neutral, merely presenting the numbers in a "just the facts, ma'am" fashion. Yet demographers *constitute* the truth about populations at the same time that they are measuring them. That is, embedded within the science are a whole set of assumptions about the world and its inhabitants that shape the sorts of questions asked and knowledge produced in the first place. As sociologist Nancy Riley and health policy analyst James McCarthy suggest, demography "does its work in a world where numbers and statistics represent modernity, science and knowledge" (2003:79). For example, demography focuses on "overpopulation" as a problem, reflecting Western attitudes toward the developing world. Also, while there has been no scarcity of demographic work about women, principally due to a heavy focus on fertility rates, demographers have often neglected examining structural issues of gender related to inequality and the social arrangements of the sexes. Riley and McCarthy write, "In the relatively narrow focus on fertility, we [demographers] often pull out pieces of women's lives, removed from the context and the landscape; in that process, we lose much of the meaning of the pieces of lives, and the aids we need to interpret our findings" (117).

Moreover, demographers' scope of problems has been limited to those requiring intervention, largely because of policy and sponsor needs. Anthropologist Susan Greenhalgh argues: "The major demand for [demography's] intellectual products . . . has historically come from governments and . . . foundations and international bodies, all with action agendas. Thus, to gain access to funds for its work, markets for its products, and even data to analyze . . . demography has had to operate primarily as a policy-relevant field, tailoring its work, and its theories, to the needs of its clientele's agenda" (1996:31). This has meant, in part, that there has been no extensive theoretical development in demography (hence the emergence of critical geography). Obviously, as feminist scholars of science have shown, epistemology (ways of knowing) and methodology (tools for knowing) are interrelated. At issue, according to Riley and McCarthy, is not just that demography relies almost exclusively on quantitative methodology: "It is the way that this methodology reflects and in turn reinforces particular epistemological directions and silences that is important." (2003:80).

In a study of declining Italian fertility rates, anthropologist Elizabeth Krause suggests that we interrogate population science for what it conceals as much as for what it reveals: "By unveiling key epistemologies of demographic practice, I put into question population-science strategies. I am convinced that the project I undertake here is a necessary one if we are to understand the dire consequences of knowledge production that masquerades as neutral science and hence as 'truth'" (2001:578). She tracks the translation of demographic knowledge into policy, focusing on the Italian fascist campaigns of the early 20th century that asserted the needs of the state above those of women. Not only is demography diagnostic, as in Krause's framework, but we would argue it is also *prognostic.* Scientists—and policymakers—"aggregate the outcomes of our intimate behaviors" (2001:579). Krause further suggests that demographers "manufacture fear" by providing scientific legitimacy to fertility crises, such as "alarming" increases or decreases in birthrates (2006).

A review of the recent literature in sociology and health disparities research reveals the ascendance of a demographic approach. In looking at infant mortality, we can see this developed over the past 25 years. Almost without exception, demographic studies position infant mortality as a dependent variable to be explained (e.g., by race or health care access) or, less frequently, as a causal factor (e.g., of a nation's health status) framed exclusively in quantitative terms. That is, infant mortality means something *insofar as it explains or is believed to represent something else,* something framed as more significant than actual dead babies. For example, in 1982 M. S. Boone measured the effect on infant mortality of alcoholism, smoking, low maternal weight at delivery, hypertension history, migrant status, ineffective contraception, prenatal care, violence, and social support, arguing that a subset of "very high risk" women produce vulnerable babies. Later, Robert Hummer et al. (1992) examined infant mortality in

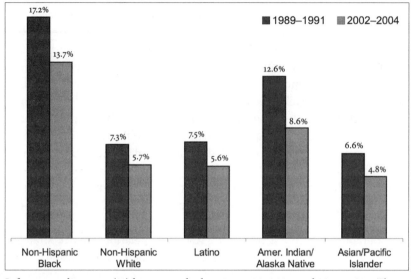

Infant mortality rates (%) by race and ethnicity, 1989–1991 and 2002–2004. This chart offers a good example of the absence of actual dead bodies in demographic representations. Source: Rogelio Saenz, "The Growing Color Divide in U.S. Infant Mortality," Population Reference Bureau, October 2007.

Hispanic populations in Florida, using variables such as maternal marital status, prior fetal deaths, timing of prenatal care, and birth order. More recently, Jamie Doyle et al. (2003) assessed the validity of Apgar scores, a standard measure of newborn health at delivery, for predicting infant survival. In all these studies, the IMR is defined as an objective fact to be measured and explained, and the bodies of infants are not discussed.

In other words, demographers perform a type of numerical magic, erasing actual bodies and the grief accompanying their loss and replacing them with statistics. Individual deaths taken together are transformed into rates, which then become the basis for implementation of public schemes. The social problem of infants dying before they reach their first birthday, a problem caused in large part by structural inequality, is displaced by the technical, and fixable, problem of a high rate. Dead bodies are translated into charts and tables. The death of a baby, and then another baby, and then another baby, some two million each year around the world, becomes in aggregate the infant mortality rate, and the IMR is the abstract language of states and bureaucrats.

Policies are instituted not in response to the bodies, then, but to the rates. The Sheppard-Towner Act and its biopolitical arrangements in the early 20th century successfully reduced the IMR in the United States, but then the government stopped worrying about infant death. The rate improved over a century, yet many of the underlying structural conditions—particularly race and its vile kin, racism—remained in place. Now that the IMR is rising in the United States, the media and government are beginning to notice. But the fundamental issue, we argue, is that *social structure is hidden inside the rate;* statistics do not merely represent inequalities, they actually shape them through rational calculation and chronic erasure of bodies and lives. We shall see next how demographic data are being used to forge new and potentially dangerous reproductive practices aimed at optimizing (fetal) health and life at women's expense.

Dubious Biopolitics of Preconception Care

We have suggested that states respond to rates, not to bodies. Demography, like all human sciences, is shaped by politics. Through the techniques of investigation and recording, the problem of dead babies is rationalized and transmuted into statistics. So, too, is the solution. To paraphrase early-20th-century Chicago School sociologists W. I. Thomas and Dorothy Swaine Thomas (1970), problems defined as aggregate are aggregate in their consequences.[8] That is, aggregate problems seek aggregate solutions. Yet a funny thing happens on the way to the morgue. The aggregate, in our view, does not necessarily translate into recognition of the structural. Solutions targeting troublesome demographic measures such as an elevated IMR may in the end reduce the rate, but they may not transform the underlying structural conditions that produced the high IMR in the first place. To return to the Sheppard-Towner Act, infant mortality rates were significantly improved over time, while the status of women and of African Americans remained unchanged for decades. Moreover, an aggregate solution may target individual behavior, not social systems, thus replicating stratification.

In April 2006, the Centers for Disease Control and Prevention (CDC) released national recommendations for a new set of preventive and clinical practices, termed preconception care (PCC), to improve maternal and child health. Clearly motivated by increasing mortality rates in the United States since 2002, the agency noted the country's low ranking among developed nations. CDC director Julie Gerberding stated, "The child-bearing years are an exciting time in a woman's life and there are a number of steps they can take to be healthy, benefiting both them and their future child," and another CDC official remarked, "Preconception health is important for every woman capable of having a baby, and should be tailored to each individual" (CDC 2006). According to the CDC, the recommendations resulted from a two-year collaboration with partner groups, including nongovernmental and health organizations, and are designed to assist health care professionals in deciding how and when to intervene.

PCC is a set of guidelines defined by proponents "as a critical component of health care for women of reproductive age. The main goal of preconception care is to provide health promotion, screening, and interventions for women of reproductive age to reduce risk factors that might affect future pregnancies" (Johnson et al. 2006:3). Additional goals include improving knowledge and attitudes of men and women related to preconception health, assuring that all women of childbearing age receive PCC services, reducing risks through various interventions, and reducing disparities in adverse pregnancy outcomes. The CDC and its partners portray this new and improved approach to child health as the gold standard of maternal-fetal care in public health and medicine, and it has been widely disseminated and adopted.

There are ten major components of PCC: individual responsibility across the lifespan, consumer awareness, preventive visits, interventions for identified risks, interconception care, prepregnancy checkup, health insurance coverage for women with low incomes, public health programs and strategies, research, and monitoring improvements. Moving between and among individual, familial, community, institutional, and state levels, the recommendations are designed to embody both an ecological perspective and a lifespan approach. The target population is defined as women, "from menarche to menopause, who are capable of having children, *even if they do not intend to conceive*" (Johnson et al. 2006:9; emphasis added).

While a full investigation of PCC is beyond our scope here,[9] we want to focus on two important consequences. First, Johnson et al. state, "The purpose of preconception care is to improve the health of each woman before any pregnancy and thereby affect the future health of the woman, her child, and her family" (2006:16). That is, the subject and object of PCC is any woman who may be biologically capable of becoming pregnant even if she has no plan to actually become pregnant. Not only does this approach significantly increase the chances that a woman will be compelled to interact with the medical system, it also greatly adds to women's responsibilities for reproduction. PCC frames *all* women in reproductive terms from onset of menstruation, no matter their actual desires and goals regarding motherhood (including the decision *not* to become a mother). In other words, girls and women are always and everywhere seen as potentially pregnant and, according to the PCC guidelines, should be treated accordingly. In this framing, women's health is defined in ever-narrower terms as maternal health—or maternal-child health—and is clearly understood to be in the service of enhancing "the quality of health for families and the community" (Johnson et al. 2006:16).

Consider the first PCC recommendation: increasing responsibility across the lifespan. According to the CDC, "each woman, man, and couple should be encouraged to have a reproductive life plan" (Johnson et al. 2006:9). Men are mentioned, yet discussion of risks—of unintended pregnancy, age-related infertility, and fetal exposures—apply primarily to women. In an article tellingly titled, "Improving Women's Health for the Sake of Our Children," Parker et al. define PCC as "a set of interventions that aim to identify and modify biomedical, behavioral, and social risks to a *woman's* health or pregnancy outcome" (2006:475; emphasis added). PCC is seen to offer numerous benefits to women and especially to their children; for women, the opportunity to lose weight and stop smoking is seen as desirable. So while the CDC's recommendations are at least on the surface geared toward women *and* men, it is principally *women* who are targeted for amplified surveillance in the name of improved fetal, infant, and child health.

To borrow anthropologist Aiwa Ong and political scientist Stephen Collier's (2005) term, PCC is a "global assemblage" of techniques, discourses, practices, and people. It is both a discursive embodiment of biopower and an object of other forms of knowledge, including demography. PCC encapsulates the sort of disciplinary corporeal practices described by Foucault, including surveillance and self-care, that operate on bodies without necessarily or actually improving health. With PCC, demography and governmentality meet at the nexus of fertility and mortality. This new assemblage brings together human actors (e.g., physicians, public health officials, nurses, ancillary health care workers, social workers, psychiatrists, genetic counselors, marketing experts, school officials, researchers, and of course women) with nonhuman actors (e.g., diagnostic technologies, pills, pharmaceutical companies, vaccines, glucose tests, statistics, reports, educational posters, insurance companies, food, government agencies, block grants, surveys, geographic information systems, computers, and registries) to address the aggregate problem of elevated IMRs.

At the same time, the CDC's recommendations significantly expand and reorganize the biomedical, public health, and socioeconomic apparatus dedicated to improving the IMR. For example, Christine E. Prue and Katherine Lyon Daniel use techniques of social marketing to consider ways to generate consumer demand for PCC: "Preconception care . . . is the product or concept we are trying to 'sell' as a precursor for assuring [preconception health]. . . . The product has been described in a number of publications as including a 'bundle' of services" (2006:S80). Kay Johnson explores the relationship between access to PCC and public financing strategies: "Without changes in financing . . . it appears unlikely that the nation will improve its preconception health and, thereby, improve the health of women, their children, and their families" (2006:85). And Grosse et al. make "the business case" for PCC: "In order to argue that preconception care is a 'good buy,' its costs and benefits must be assessed from the perspectives of a variety of stakeholders involved in health care services" (2006:94).

Numerous organizations, ranging from the American College of Obstetricians and Gynecologists to the National Association of County and City Health Officials, have jumped on the PCC bandwagon with gusto. One such organization is the March of Dimes, the nation's premier non-profit infant health organization. Long an advocate for the prevention and health of premature babies and a sponsor of and clearinghouse for research on birth defects, the March of Dimes has broadened its public health message to include PCC and produced a slick promotional campaign. The organization offers a free online PCC curriculum for physicians (including questions to help girls and women decide if they want to become mothers), tools for risk reduction (such as recommendations for folic acid, diagnostic tests, and drugs, smoking cessation, weight loss, and vaccination), information about genetic counseling and financial planning, and a list of ten things to do *before* pregnancy.[10] These include taking folic acid, avoiding cigarette smoke and alcohol, preventing infections, avoiding hazardous substances and chemicals, reducing or avoiding stress, and getting a pre-pregnancy checkup.

What we find most striking about the CDC recommendations and their interpretation by the March of Dimes and other organizations is the near absence of representations of actual infant death. As framed statistically, the IMR is a measure, a demographic category—no more and no less. PCC is avowedly motivated, in part, by rising infant mortality rates in the United States. Yet neither the CDC recommendations nor public education materials such as those produced by the March of Dimes frame infant mortality in terms of death, grief, and bereavement, nor do the

EASY TO READ

Take Folic Acid
Tome ácido fólico

Use this easy-to-read flyer to remind women of childbearing age to take a multivitamin with folic acid every day, as part of a healthy diet, to help prevent neural tube defects. (Learn more about our easy-to-read materials on page 16.) *New English version available Summer 2006.*

09-1103-98 *Available while supplies last.*
Flyer – English $6.00/pkg 50

09-1104-98
Flyer – Spanish $6.00/pkg 50

Think Ahead for a Healthy Baby
Diez consejos para tener un bebé saludable

Offers ten simple steps a woman can take before pregnancy to prepare for a healthy baby. Includes tips on taking folic acid; the hazards of drinking alcohol, smoking and drugs; and the importance of a pre-pregnancy checkup. *Available while supplies last.*

09-823-95
Pamphlet – English $12.00/pkg 50

09-1215-99
Pamphlet – Spanish $12.00/pkg 50

Co-branding
is available for an additional charge.
Discounts available for bulk orders.
Call 1-914-997-4558 for more information.

Folic Acid is Good for Me/Folic Acid is Good for Us
El ácido fólico me hace bien/El ácido fólico nos hace bien

These uniquely designed pamphlets encourage women of childbearing age to take folic acid even if they are not yet ready to have children. Explains the various ways in which folic acid promotes good health before, during and after pregnancy.

53-1657-02
Pamphlet – English
(African-American Cover) $6.00/pkg 50

53-1658-02
Pamphlet – English (Caucasian Cover) $6.00/pkg 50

53-1483-00
Pamphlet – Spanish $6.00/pkg 50

A page from a March of Dimes catalog featuring preconception care (PCC) resources.

bodies of infants (alive or dead) appear in these accounts. Just as demographic studies position infant death as a statistical problem, so do these recommendations abstract the actual demise of babies into preventive guidelines and procedures. Death in the aggregate, the calculable problem of infant mortality, is met not with mourning but with a biomedical apparatus: a set of disciplinary procedures amenable to widespread implementation and "systematic review."

Digging more deeply into the amassed demographic data on human infant death and policy solutions such as PCC, we can see that the infant mortality rate is laid squarely on women's and girls' shoulders, and thus that is where the fate of the nation's health rests. It is *women's* (not men's) bodies, behaviors, habits, employment, relationships, choices, and practices that are seen to determine whether their babies will live or die. We want to point to perhaps the largest irony of all: we use aggregate data to measure and create problems through demographic truths, then fail to pursue *truly* aggregate (i.e., structural) solutions such as expanded health care access or elimination of poverty and racism. In current articulations of U.S. government policy, infant mortality may be a growing social problem, but the solutions target women's *individual* bodies and practices. Babies may be invisible (and thus innocent) in these frameworks, but women are highly visible containers of blame.

"It's Like Losing a Part of Your Heart"

What might discussions of infant mortality look like if we brought missing bodies back in? What sorts of experiences are made visible by thinking through the bodies of dead infants? How might we humanize the IMR? It seems to us that narratives about loss might be one place to start. The media, for example, does attempt to cast the issue in emotional terms, framing infant death as spectacle, trauma, and loss in an effort to reach readers and, we suspect, sell newspapers. Consider this passage from a recent story about the very high IMR in Memphis, Tennessee, a Southern city with a significant African American population (63.5%) compared with the white population (30.5%) (U.S. Census Bureau 2006): "A pickup truck and a backhoe show up on the days, usually Tuesdays and Thursdays with good weather, when babies are buried at the county cemetery. The first carries the little wooden coffins, and the second digs the hole, maybe three feet wide, where they are placed a foot apart" (McClam 2007). In the same article, a nurse describes dying infants in the neonatology ward as

The Bearly Gone Keepsake Angel Urn, available in 2008 from www.holyurns.com for $95.95. A wide variety of infant caskets and urns are available from the funeral industry.

"too small and too sick to cry." This is the IMR rendered in local, graphic, heartbreaking terms.

In 2002, an award-winning article by *Baltimore Sun* journalist Diana Sugg explored stillbirth. It opened with this description: "That chilly night in late October, the delivery room was so quiet. The doctor wrapped the 8-pound, 21-inch newborn girl in a pink-and-blue striped cotton blanket, pulled a matching cap over her brown hair and gently passed her to her mother. Margarete Heber cradled the baby. In the dim light, Heber could see the infant had her dark eyes, turned-up nose and distinctive chin. Perfect, except she was tinged with blue. She had died just hours before she was born. Her birth would be her good-bye." After this gripping, emotional paragraph, the story went on to describe scientist Heber's search for an answer to the mystery of why her daughter, Elisabetha, died before birth. She discovered that there are more questions than answers regarding the problem of stillborn babies.

We are sympathetic to the quest for answers to the problem of still-birth—which results in almost as many deaths *before* birth, some 26,000 annually, as die in the first year of life. Yet we want to highlight here Sugg's discursive embodiment of dead babies not only in the description of Elisabetha's death but throughout the article. She discusses the "special gravesites" created by three local hospitals. She quotes a woman whose baby died seven years previously: "That baby was real to me the moment I knew he was there. It's like losing a part of your heart." She describes hor-rific practices of previous generations: "nurses hustled stillborns out of the delivery room without showing them to the mothers. Hospitals disposed of most of the babies as pathological specimens." Sugg chronicles the ex-periences of a woman whose baby stopped moving after a typical Lamaze class: "Early the next morning, doctors gave her drugs to deliver the baby, and during the two full days of labor, Norton and her husband had to decide on an autopsy and funeral arrangements for the child they hadn't seen or touched. At 10:30 p.m. on Aug. 27, Liz Norton delivered a boy. His father held him and recognized his own features in his son's face."

Aside from media accounts, such poignant and descriptive accounts also appear on the many Internet sites dedicated to infant loss through miscarriage, stillbirth, sudden infant death syndrome (SIDS), and other death. For example, the site My Magical Memories details the death of a newborn: "I went to check on MacKenzie, who was asleep in our room. All was normal. As she came into view, I realized all was no longer nor-mal. . . . Our daughter, our baby, did not live. She died that morning, in our home, just feet from me. I was thrown into a life I had not planned and believed only happened to other people. I was now faced with plan-ning a funeral for my daughter and having to make decisions regarding things way beyond my imagination: cremation or burial, open casket or closed, the list goes on."[11] The author decides that she must continue to live, despite her grief, because of her duty to her remaining children. Yet she vows to live her life in memory of MacKenzie.

On the Mommies Enduring Neonatal Death (MEND) blog, a Chris-tian support network, letter after letter embodies the suffering of parents who have lost an infant. One woman writes, "The four-month mark . . . I began to forget what she felt like in my arms and what she smelled like.

People began to move on with their lives and forget the tragedy I'd been through." Another woman laments in a poem, "The hole in my heart that can only be filled by you . . . invisible. The tears that I shed as I, alone, hang your ornaments on your siblings' Christmas tree . . . invisible. The scar tissue that has formed around my heart and gets thicker as the years go by . . . invisible . . ." And the family of Rebekah writes, "It's so hard to believe that it has been four years since we saw your sweet face and met you for the first time. I can still remember it as if it were only yesterday."[12]

Letters to Heaven, a now-defunct searchable Internet site, featured personalized greetings and remembrances "sent" to infants who have died. A grandmother wrote to her dead grandson, "My heart is so empty without you, I just want to hug you and hold you close to me." A mother wrote to her daughter, "You're supposed to be with me right now climbing up on my lap and pulling my keyboard down and yelling at me for attention. I'm supposed to have all three of my girls crawling all over me and demanding 100 percent of my attention. I often ask why I have to live the rest of my life knowing that on earth I can't hold you, kiss you, or smell your sweet breath. I want so badly to smell you, especially your little feet." Another mother wrote to her son, " I wonder who you look like now, I am sure you are as handsome as your Daddy, but still have your Mommy's eyes and laugh." And a grandmother wrote to her Little Pito, "I really want to squeeze you with a strong hug. I know you would squirm to get away because you would be tired of me crowding you, but that's just the kind of abuela I am."[13]

The nonprofit organization First Candle, dedicated to advancing infant survival, captures these embodied experiences of grieving parents in its booklet *Surviving the Death of a Baby*. The organization acknowledges that the death of an infant may be the most painful experience of an adult's life and describes a grieving process characterized by shock, sadness, guilt, anger, and fear. Yet it also notes corporeal symptoms of grief, including inability to concentrate, dizziness, pressure in the head, and exhaustion: "Those in grief often experience muscular problems or other physical symptoms centering around the heart or stomach. Often they have no appetite, and they eat only because they know they must. They feel 'tied in knots' inside. Parents often say that their arms 'ache' to hold the baby." Mothers whose milk has come in face a constant, agonizing, liquid reminder of their loss.[14]

In these narratives, infant mortality is anything but a statistical category or abstract measure. The numbers, it turns out, have names: Ciarra,

Bobby, MacKenzie, and Little Pito. Infant loss is clearly a deeply felt experience of intense emotional and bodily suffering, a fact that might be unknown to those who have not lost children if one were to read only demographic reports about the phenomenon of "rising IMRs." Among surviving parents, hearts break, stomachs churn, heads split, tears spill. Mothers, especially, describe the loss of a child in terms of their own bodies; their arms feel empty, their stomachs are hollow, their breasts are sore for weeks, sometimes years. Some women experience infant loss in the same way that an amputee suffers from Merleau-Ponty's phantom limb (Harrison and Kositsky 1983). Thus, loss of a baby (the "limb") is akin to the loss of a part of the self, and there are chronic, unending reminders of its absence.

Sociologist Patricia Ticineto Clough asks, "What is the ontological status of a ghosted body, of a haunted materiality?" (2007:7). What most strikes us about the subjective and profoundly intimate love letters discussed above is how much these dead babies are still longed for. They are, indeed, missing bodies. Because these narratives contain within them such revealing portraits of lost children and embodied suffering, they elicit a level of affect absent from actuarial representations of the IMR. In place of rationality, we get emotionality. We can almost see and feel the flesh of the missing babies, those desired little creatures doomed by their materiality. They can live now only in words and remembrances. Yet we also worry that maternal narratives of loss are suspect; women who lose children are perhaps seen as too close to death, too emotional to provide an "accurate and objective" account of infant mortality. Indeed, evaluated by emergent biopolitical standards, these women would be found wanting; their babies died, after all. The reproductive experiences of these women are counted, unfairly, as failures of modernity.

Registering What Counts

We have traversed in this chapter from aggregate data and numbers to actual flesh-to-flesh relationships, from the disembodied to the intensely embodied, from numbers to people, from measuring death to narrating loss. We are not suggesting here that numbers do not matter. They do, in particularly acute ways both for governments and for feminist scholars engaged in freedom projects. But centuries of statistics and a demographic imperative to count have not, in fact, prevented a recent increase in the IMR for certain groups of people in the United States. Neither has dry recitation and deployment of statistics managed to capture the public's imagination in the service of social change. We should not simply do away with numbers: quantification is persuasive and makes important connections between and among variables. Yet because numeric data often conceal social structures (e.g., hiding dead bodies in a table or graph), they also shape policy and biomedicine in narrow ways that can place potentially quite restrictive burdens of surveillance and behavioral modification on women. As we have seen with PCC, women's bodies are the not-so-innocent, fertile terrain on which solutions to infant death are manifest and women's own grief is expunged.

We want to return our attention to the penguins, those charming, fuzzy critters whose arduous tale inspired glee, hope, familial connection, cross-species understanding, and perhaps a commitment to environmentalism. The downy bodies of the vulnerable baby penguins haunt us long after we leave the movie theater. We are moved to action: we take Al Gore's Nobel Prize–winning message of climate change to heart, and we want to save the planet for the sake of the penguin babies (and the polar bear, whale, and seal babies, too). The penguins are part of our cultural imaginary in a way that human babies, especially dead babies, are not. (We were intrigued to learn that penguins can now be biometrically monitored, assuring them a place in our surveillance society [Morelle 2008].)

Is the loss of human babies so huge and agonizing that we are simply incapable of registering it as something other than the IMR? Do we cling to a kind of collective denial that our own species is fragile and embattled on a precarious planet? Is it just easier to blame individual women—especially poor women of color or third-world women—and their supposedly faulty behaviors and bodies for deeper structural problems? Is it more conducive to governments to hide racism, misogyny, and fear inside a demographic register, with its ostensibly apolitical numbers, or inside biomedicalized practices such as PCC, than it is to care enough about women to *empower* rather than belittle and control them in the service of healthier babies?

In posing these questions, provoked by our analysis in this chapter, we want to suggest the inclusion of a different register, one peopled by actual bodies living and dead. Suffering and ill children, abject grief and unbearable pain, material practices such as infant burial—all can help to expand our gaze to encompass dead babies and those who mourn them. Drawing a lesson from the penguins, if we can visualize human parental struggles to help babies thrive in circumstances of deprivation or witness the loss of beloved babies, then perhaps the seeds of collective action may be sown. As authors and mothers, we obviously want to bring babies back into the frame, and yet we have quite limited ways of doing so. To actually picture the bodies of dead babies feels risky, insensitive, and gratuitous. The phrase "dead babies" itself is jarring, yet we use it intentionally here to help make them visible. It is painful enough to read about the loss of these babies on the seemingly private but terribly public Internet sites dedicated to infant death. Just as loss is relegated to the hidden, somewhat shameful recesses of the family, our account, too, is incapable of fully visualizing or documenting the loss of infants. While a proliferation of images of the dead children of our human neighbors might move us to care and act, at the same time, such images would surely be seen as unpalatable. Perhaps it is far easier to blame women and to direct their grief into the "proper channels" than it is to collectively bear witness to unnecessary species losses for which we are *all* ultimately responsible.

Unfortunately, as we show in the next chapter, simply representing death and loss is not enough to inspire government action that is quantitatively *and* qualitatively sufficient to help people. While images of suffering may move individuals to care, such visual evidence does not automatically translate into official policies beneficial to the aggrieved. As we have argued here, biopolitics of various kinds—especially demographic calculations—shape the solutions prescribed for social problems defined in the aggregate, reproducing inequalities in the name of data. In the next chapter, we see in HIV/AIDS another example of how quantified objectification of disease and death leads to the erasure of embodied suffering. Visibility of bodies matters, to be sure, but so do the obdurate realities of gender, race, class, and geography.

PART II

Exposed

> Life is either a daring adventure or nothing. Security does not exist in nature, nor do the children of men as a whole experience it. Avoiding danger is no safer in the long run than exposure.
>
> Helen Keller, *The Open Door* (1957)

Becoming a visible body, a body that counts and is taken seriously, involves the experience of being seen by a critical mass of people with power and institutions of power. Exposure, the degree of coverage and the amount of attention the body receives, is one way to conceptualize the process of bodies coming into view. Exposure is always sensational: baring the body and its fleshy parts to the elements and rendering the unshielded body at its most vulnerable. Yet becoming exposed is paradoxical. Just as being exposed can be dangerous or damaging to bodies, it also can bring bodies into focus and mobilize resources to consider these bodies and their unique situations. Individuals can be exposed as fraudulent, or they may die of exposure. People may be accused of indecent exposure, or they may be overexposed.

Sociology instructs us to consider the conditions and circumstances of the social production of exposure. What social order is maintained by the unequal distribution of exposure? Who benefits from exposure, and what types of exposure are most advantageous? Under what conditions is someone exposed *to* something or *as* something? The dimensions of quantity and quality of exposure include the amount of time a body is exposed, to what it is exposed, the body's susceptibility to exposure, experiences of exposure, and consequences of exposure. There are widely varying degrees of consent with which individuals participate in their exposure. We suggest comparing three uses of the term exposure that we find analytically useful.

In photography, exposure is the amount of light that falls on the film in the process of taking a picture. But, as anyone knows who has forgotten to set the flash or has opened the back of a camera, film can be underexposed or overexposed and images lost forever. Degrees of exposure matter in achieving the desired effects. Something comes into view, we adjust the aperture, and the image is captured by our camera, kept in digital memory.

In journalism, an exposé is an investigation into a situation that intends to reveal shocking, previously hidden, or surprising information. In the tradition of Upton Sinclair's *The Jungle* and muckraking newspaper coverage, reporters go "under cover" to expose or reveal a scandal or to make public the operations of power, inequality, transgressions, or greed. More recently in the West, exposés have broadened their lens beyond social problems and issues to disclose the secret lives of famous people, including their behavior, habits, relationships, body problems, sexuality, foibles, and philanthropic activities.

In epidemiology, exposure is a conceptual and methodological category used to compare dichotomous variables of the exposed and the unexposed in relations between human beings and their environments. Most often, exposure of humans occurs through contact with the skin, ingestion through the mouth, inhalation through the nose, or exchange of bodily fluids. Epidemiologists are trained to measure air, soil, water, and food for toxins and pathogens. Thus, exposure illustrates our interconnection with each other and with the planet. As such, exposure has become a critical term in the environmental justice movement and environmental health literature. As we mourn the loss of the unsoiled planet, we simultaneously construct a "memory"—a factoid—of when the earth and our bodies were clean, before exposure. The unexposed retain a sense of innocence, as they are uncorrupted by forces that might present danger or weaken their defenses. And the exposed are often blamed for their exposure, even if it derived from circumstances beyond their control. Just as exposure can bring bodies into view, it can also poison and kill them.

Ironically, exposure can also sometimes be considered protective. Exposure to fluoridated water as a child is protective against cavities and gum disease later in life. Exposure to the sun is necessary for vitamin D absorption, but overexposure can lead to deadly skin cancer. A delicate balance, then, must be negotiated between too much exposure and not enough.

Bodies must be exposed in order to be seen and, consequently, longed for. However, there is an unequal distribution of exposure to danger, risk, and disease; and because certain bodies do not garner attention or visibility, they are often missed.

4

Biodisaster

"The Greatest Weapon of Mass Destruction on Earth"

The tragic truth is that until some kind of actual cure is discovered, most people with HIV/AIDS in the developing world are essentially doomed.

Nicholas Eberstadt, "The Future of AIDS" (2002)

In February 2008, President George W. Bush visited Africa, where he announced, "My trip here is a way to remind future presidents and future Congresses that it is in the national interest and in the moral interests of the United States of America to help people." Pronouncing the idea of democracy through Benin, Rwanda, Ghana, Liberia, and Tanzania, Bush wanted to highlight his administration's commitment to funding for AIDS, malaria, and education efforts and also likely wanted to take the media focus off the disastrous war in Iraq. These programs in Africa, financed in part through the Millennium Challenge Corporation, which supports only nations that make a commitment to democracy and free markets,[1] were seen as key elements in Bush's government of self-proclaimed "compassionate conservatism." He left the more difficult venues—incendiary Kenya and strife-ridden Darfur—to his secretary of State, Condoleezza Rice, who framed the visit in terms of political damage control. Responding to media criticism of his refusal to visit the troubled zones, Bush stated, "This is a large place with a lot of nations, and no question not everything is perfect. On the other hand, there's a lot of great success stories, and the United States is pleased to be involved with those success stories."[2]

Taking these comments at face value, one might think that the 2008 visit to Africa involved parallel but distinct paths, one along humanitarian lines and one along vectors of global security. While the president was putting a human (and American) face on efforts to combat disease and poverty, Rice was applying her unique "neo-con" brand of political muscle to the dangerous and bleak situations in Kenya and Darfur. Yet it would be more accurate to perceive the Bush Administration's visit to Africa as a strategic, multi-pronged approach to "securing" an embattled continent. In presenting both the "soft" and "hard" arms of U.S. foreign policy, Bush and his team indicated clearly that Africa was firmly on their radar as a security concern. This approach was entirely consistent with U.S. public health policy since the 1990s, and also with the 2000 U.N. Security Council Resolution "warning that the HIV/AIDS pandemic, if unchecked, could threaten world stability and security" (Garrett 2005a:51). But it differs radically from remarks made by George Bush in February 2000, in an interview conducted before he was elected president: "While Africa may be important, it doesn't fit into the national strategic interests, as far as I can see them."[3]

This chapter tracks a shift in the conceptualization of HIV/AIDS— namely, its emergence as an issue of national and global security. Unlike in the 1980s, when it began to be framed in terms of health, disease, and epidemiology, in the 21st century it is considered to be central to the security interests of nation states. While HIV/AIDS is still "the most complex disease humanity has ever faced," as prize-winning journalist Laurie Garrett terms it, it may also be responsible for "new threats to stability and security . . . as the pandemic escalates" (2005a:52). The United States, especially, tends to view such threats through a lens sharpened by the experiences of 9/11, or what Garrett calls "the prism of terrorism." Reframing HIV/AIDS as a security issue promises to bolster support and redirect funding to combat the disease. While 26 million people have been killed by AIDS (more than 0.5 million in the United States) and another 40 million worldwide are afflicted—shocking numbers by any epidemiological or moral calculation—global efforts at prevention and eradication have been remarkably anemic. We are concerned here with particular representations of human bodies in these new configurations of disease as objects of foreign policy and with altered definitions of risk. Specifically, if HIV/ AIDS is a security issue, then who is threatened with exposure? Is it the individuals and communities decimated by disease, or is it the "rest of us" in healthy, prosperous nations who may be negatively affected by states destabilized by structural adjustment, public health crises, and longstanding ethnic conflicts exacerbated by foreign intervention?

Everyday we in the West are bombarded with sensational and dramatic information about health. The ubiquitous cable news scroll informs us of X disease affecting Y group of people at unprecedented rates or of outbreaks of strange symptoms targeting a geographic area or demographic cohort. Newspapers feature stories about drug discoveries and amelioration of suffering one day, and the next tell us about the unforeseen side effects of pharmaceuticals. Prescription drug advertisements depict happy, healthy (white) people in soft focus, clearly enjoying their lives, as a rushed voiceover admits that side effects might include anal leakage or liver failure. There is a certain balance struck between the constant feed of "information" about health and its fragile achievement, and the ever-present risks to it. Media coverage routinely profiles biomedical innovations and health surveillance techniques that purport to secure our claim to fleeting health. Yet we are kept in a constant state of insecurity about our bodies, and our vulnerability to emerging threats is amplified.

HIV/AIDS statistics are commonly used in this fashion. Twenty-five years ago, HIV/AIDS was a public health crisis that everyone discussed. It was a frightening epidemic that, after initial U.S. inaction, generated changes in activism, public health practices, clinical research, funding streams, and sexual politics (Patton 1990, Treichler 1999). Once the disease was brought "under control" in the United States (a myth), HIV/AIDS dropped off the radar of the media and policymakers. Whereas in the 1980s and early 1990s, there was at least one HIV/AIDS story a day in major newspapers, at the turn of the century the disease was rarely discussed. What is so shocking about this is that HIV/AIDS is now a more complicated and dangerous disease than it ever was, affecting millions of people around the globe. No longer considered merely an epidemic, NGOs and public health experts frame it as a global *pandemic,* and it is endemic to certain communities especially in the developing world. Geography matters: as HIV/AIDS has "migrated" from the United States "back" to Africa and to Russia, India, China, and other non-Western spaces, most Americans have grown increasingly less concerned about the disease. It is something that happens to "other" people "over there," and not here—tragic, to be sure, but not "our" concern. In the United States, there are few memorial quilts for dead Africans (just as there were few for dead African Americans).

It might seem odd to include an analysis of AIDS in a book called *Missing Bodies.* After all, the vulnerability of the body is obviously displayed in HIV/AIDS discourse, saturated as it is with historical images of decaying human beings on a painful road to eventual death, slowed but not stopped by antiretroviral therapy (ART). The faces of AIDS—young and old, gay and straight, "virtuous" and "deviant"—have been with us for more than two decades. However, within the contemporary United States, our constructions of endemic HIV/AIDS, now discussed narrowly in the shorthand of "African AIDS," have moved away from the visibility of afflicted American bodies to the disembodied surveillance of foreign populations and states. We are a nation focused much less on T-cells than on proliferating terrorist cells. People with AIDS seem like a relic of the past, a sort of quaint public health concept from the previous century, much like consumption from the Victorian era, irrelevant in the post-9/11 world where identities have been fractured and stitched back together in new digitized ways. No longer are we at risk of exposure to HIV, the story goes; now our very way of life is threatened by instability of developing nations conceptualized as mutant viruses. Little matter that we are still at risk of exposure to the disease, and rates of infection in the United States are rising, especially among women and African Americans. Our attention is drawn instead to the security threats posed by rogue nations in various states of biological and political disintegration.

This is markedly different from the public health and humanitarian response of the 1980s and 1990s. Living and working in San Francisco throughout the 1990s, we witnessed a proliferation of community-based organizations, public health practices, and civic programs—a thriving social movement—designed to help people with AIDS and to mobilize local citizens to do their part. For example, the NAMES Project created the AIDS Memorial Quilt, Project Open Hand provided nutritional assistance, Shanti offered wellness services to the afflicted, PALS and PAWS cared for the pets of people with AIDS, the STOP AIDS Project offered HIV testing and medical care, and the Black Coalition on AIDS provided culturally sensitive services to African Americans affected by the disease. All these organizations still exist, but a certain fatigue has set in and challenged the sense of solidarity that they may have provided to the community and to national visibility.

While San Francisco is still a city profoundly affected by HIV/AIDS, there is an especially palpable sense of loss in regard to the disease, which is no longer nearly as present in public spaces. Urban settings have been redefined as less about public health risks and more about the threats to national borders. Immigrants, terrorists, and WMDs, rather than viruses, are what we fear and what compels civic action. Lisa lives in New York,

Poster on the side of a New York City MTA bus proclaiming the success of the "see something, say something" campaign. Source: Gur Tsabar, publisher of GurBlogs and Room Eight.

where the subway is filled with posters and piped-in messages urging riders, "If you see something, say something" to the police, who now regularly patrol the platforms—potent reminders of the post-9/11 security apparatus. And Monica lived in the evangelical South ("the buckle of the Bible belt"), where HIV/AIDS presumably does not exist, if one judges by media and billboard attention, but ever-present threats to unborn fetuses and God do.

In narrating a transformation in how the United States has conceptualized and responded to HIV/AIDS—from public health crisis to global security breach—we offer a critical reading of epidemiology as a biopolitical discourse. We show how certain statistical measures for evaluating bodies are co-constitutive with political economic concerns. By locating risk and emerging threats in standardized and universal numeric language, epidemiology presents information about HIV/AIDS in ways that make it portable across geographic borders. While the specific techniques of epidemiology have been largely consistent since the emergence of HIV/AIDS as a crisis for state management, the meanings of the disease have changed. In the early days of the epidemic, numbers were used to translate risk and disease into a clear and present public health danger. In the present moment, in which issues of biosecurity are ascendant, risk is translated in terms of disembodied arms control. Whereas suffering and infected bodies were visible in the public health era, in the post-9/11 security era bodies are increasingly invisible to us. In the developed world, people with AIDS are concealed by pharmaceuticals which render infection undetectable, and in the developing world they are dehumanized and framed routinely as stealth weapons of mass destruction.

Public Health, Epidemiology, and Politics

The field of public health has been marked as beginning with rudimentary measures taken of suffering and mortality caused by plague during the Black Death of the 14th century. Commonly today, the public health era is considered to be the late 19th and early 20th centuries, which saw the rise of sanitation and other forms of social hygiene. Scientific tools such as quantitative methods and statistical models were enrolled in the pursuit of a more accurate and policy-relevant public health practice. The latter half of the 20th century can be characterized in terms of a population health model, in which measuring the distribution of health and sickness helped to create better understandings of health disparities. Both public and private funds were largely targeted to "outbreaks"—acute, dramatic, and potentially lethal conditions such as hepatitis C—rather than to chronic, silent epidemics such as depression or arthritis. (The "war on cancer" remains a notable exception.) Pursuit of the causes of disease is typically considered much more exciting and heroic—the public health officer as gumshoe—than implementing preventive programs. The defunding and collapse of public health infrastructure in the United States paved the way for emergence of a new public health model, one defined by biopolitics of monitoring and surveillance. Yet even within this transformed approach, numbers matter.

In what follows, we both critique and rely on numeric accounting. That is, any contemporary understanding of HIV/AIDS requires empiricism, or estimations and calculations of people sick and dying. Numbers are powerful symbols of the magnitude and scope of the crisis. Yet, as earlier suggested with our focus on demography, at the same time, we assert that reducing all phenomena, perhaps especially human bodies and lives, to *mere* numbers is an act of violence. Relying on quantitative assessment helps us figure out how to manage a public health crisis such as "African AIDS." But we must also recognize that epidemiology, like demography, performs a magic trick of concealing the behind-the-scenes work that goes into replacing pain and suffering with statistics. Part of evaluating these politics is critically examining how the sheer weight of the numbers of dead evokes mouth-gaping awe yet may not correspondingly lead to actual social change and betterment. The abacus of risk and exposure, by wrapping up millions of ravaged bodies inside the statistical packaging of the aggregate, hides the relentless suffering. People with AIDS are among the missing, in part because the sick, dying, and already dead are collapsed into simultaneously shocking and numbing figures, rationally graspable while also affectively just out of reach.

Epidemiology, the quantitative core of public health, is a science concerned with determining the distribution of diseases and the estimation of risk and rates of exposure among a given population. It provides methodologies and statistical formulas to measure the incidence and prevalence of diseases. Ultimately, it is a biopolitical tool for measuring and managing bodies. As with demography (discussed in chapter 3), epidemiology is entangled within the political economy of nations and corporations as a science of managing aggregate data and forecasting morbidity and mortality. The insurance industry, health care organizations, municipalities, human resource managers, and tax revenue boards all have vested interests in epidemiological truths and especially predictions. Epidemiology is thus enrolled in the operations of state power, calculating the flows of resources and the potentialities of risk.

In a world of limited resources (or, rather, inequitable distribution of resources), a variety of statistical tools are marshaled to make the case that certain diseases have higher social costs than others and thereby require greater economic and political investments. Take, for example, the epidemiological measures "years of potential life lost" and "years of life expectancy lost" (YPLL and YLEL), which are often used interchangeably. YPLL calculates the years of life a person lost had they lived their expected life span, and it is used as a means to measure the social and economic cost in lost productivity from premature death. The statistic can then be deployed to indicate the loss of tax revenues such as Social Security contributions and to provide "evidence" for the significance of a health crisis. Furthermore, the cohort affected by the health crisis that strikes them down in the prime of life is often constructed as a "lost" generation in terms of combating disease and averting the instability it causes. That is, the people who are dying are the ones who, given their social and professional integration, could have contributed to maintaining the stability of the affected community or nation had they not died. Relative to other diseases, such as cancer, that strike later in life but may be more prevalent, HIV/AIDS is a disease with greater YPLL than other diseases (Healton et al. 1992).

Navigating the epidemiological statistics and their associated acronyms is often quite challenging. In addition, epidemiology is often in the business of predicting rates—predictions that sometimes require revisions. As published in the *2007 AIDS Epidemic Update,* WHO and UNAIDS revised the estimated number of people living with HIV downward from 39.5 million in 2006 to 33.2 million in 2007. The revisions are due mainly to improved methodology, better surveillance by countries, and changes in the key epidemiological assumptions used to calculate the estimates.[4] What this indicates is that, despite the field's claim to objective scientific truth, in reality epidemiology is implicated in other disciplinary struggles, such as demography's quest for better ways to count and track people. It is also shaped by late capitalist constraints on how problems can be framed; that is, questions are determined not by epidemiologists but by the various interests they serve. States and corporations are deeply interested in the measurements YPLL and YLEL, which as objects of technical knowledge help policymakers to predict how long a citizenry will live, consume, work, and pay taxes.

As a discipline that trades in risk and forecasts potential threats, epidemiology may be thought of and indeed is deployed as a prophylactic technology. And yet because the prediction of risk is also the quantification of various behaviors into an actuarial category called "risk," average human beings can never keep up with the proliferating number of potential threats in the world (e.g., child sex predators, testicular cancer, weapons of mass destruction, and terrorism). No matter how large the numeric shield, it can never be adequate to protect us, the moving targets, from these excessive threats. Following recent sound bites based on epidemiologic surveillance, the magnitude of risk of HIV/AIDS is perplexing. Is it reassuring that in 1995 HIV/AIDS was the leading cause of death among 25–44 year olds in the United States and in 2008 it was fifth? Or should we be alarmed that some forms of HIV/AIDS are drug resistant and there is a group of men who are "superinfectors" who "bareback" (engage in intentional insertive anal intercourse without a condom)? How should we understand the oft-cited sub-Saharan "African AIDS" epidemic? And what about our own national borders? Are we "protected" from "global AIDS" as if by an invisible fence made of latex surrounding the United States?

As with most diseases, the manifestation of HIV/AIDS varies based on global and local stratification of power and resources. Approximately 68% of all people with HIV (or seropositive people) live in sub-Saharan Africa. Life expectancy in several African countries has been profoundly affected by death from AIDS. According to a 2005 U.N. report, in Swaziland it has been estimated that life expectancy at birth, which is currently just 33, would be 66 without AIDS (UNDP 2005). The impact that AIDS has had on average life expectancy is partly attributed to child mortality, as increasing numbers of babies are born with HIV infections acquired from their mothers. This type of infection is known as vertical transmission. The biggest increase in deaths, however, has been among adults between the ages of 20 and 49 years, precisely those citizens who can contribute most significantly to social cohesion. This group now accounts for about 60% of all deaths in sub-Saharan Africa, compared with 20% between 1985 and 1990, when the epidemic was still in its early stages (UNAIDS 2006). Domestically, the rate of AIDS in the United States is 13.7 per 100,000 population, with the highest rate among African Americans at 54.1, the next highest among Hispanics at 18, and Native Americans and Native Alaskans third at 7.4.[5]

Data indicate that the global ratio of women to men living with HIV has remained stable from 2001 to 2007. The estimated 15.4 million women living with HIV in 2007 was 1.6 million more than in 2001. For men, 15.4 million were estimated to be living with HIV in 2007 compared with 13.7 million in 2001. One way to read these epidemiological data is to suggest that in every geographic location around the world, men and women have equal rates of HIV infection. However, there are gender disparities by region that complicate this understanding. Nearly 61% of all adults in sub-Saharan Africa with HIV were women. This is compared with 43% of women in the Caribbean in 2007 (an increase from 37% in 2001). According to the UNAIDS *Epidemic Update* in 2007, the proportions of women compared with men living with HIV in Latin America, Asia, and Eastern Europe are slowly growing. In Eastern Europe and Central Asia, it is estimated that women accounted for 26% of adults with HIV in 2007, compared with 23% in 2001. In Asia, the proportion was 29% in 2007, compared with 26% in 2001.[6] The figure here shows the total number of persons by region living with HIV, new infections, adult prevalence, and numbers of deaths due to AIDS in 2001 and 2007.

Establishing the prevalence and incidence rates of HIV/AIDS is a political act and one that is fraught with many inconsistencies and ethical conundrums. Commonly establishing the numbers, especially in resource-

Regional HIV and AIDS Statistics, 2001 and 2007

	Adults and children living with HIV	Adults and children newly infected with HIV	Adult prevalence (%)	Adult and child deaths due to AIDS
SUB-SAHARAN AFRICA				
2007	22.5 million	1.7 million	5.0%	1.6 million
	[20.9 million–24.3 million]	[1.4 million–2.4 million]	[5.5%–6.6%]	[1.5 million–2.0 million]
2001	20.9 million	2.2 million	5.80%	1.4 million
	[19.7 million–23.6 million]	[1.7 million–2.7 million]	[5.5%–6.6%]	[1.3 million–1.9 million]
MIDDLE EAST AND NORTH AFRICA				
2007	380,000	35,000	0.30%	25,000
	[270,000–500,000]	[16,000–65,000]	[0.2%–0.4%]	[20,000–34,000]
2001	300,000	41,000	0.30%	22,000
	[220,000–400,000]	[17,000–58,000]	[0.2%–0.4%]	[11,000–39,000]
SOUTH AND SOUTH-EAST ASIA				
2007	4.0 million	340,000	0.30%	270,000
	[3.3 million–5.1 million]	[180,000–740,000]	[0.2%–0.4%]	[230,000–380,000]
2001	3.5 million	450,000	0.30%	170,000
	[2.9 million–4.5 million]	[150,000–800,000]	[0.2%–0.4%]	[120,000–220,000]
EAST ASIA				
2007	800,000	92,000	0.10%	32,000
	[620,000–960,000]	[21,000–220,000]	[<0.2%]	[28,000–49,000]
2001	420,000	77,000	0.10%	12,000
	[350,000–510,000]	[4900–130,000]	[<0.2%]	[8200–17,000]
OCEANIA				
2007	75,000	14,000	0.40%	1200
	[53,000–120,000]	[11,000–26,000]	[0.3%–0.7%]	[<500–2700]
2001	26,000	3800	0.20%	<500
	[19,000–39,000]	[3000–5600]	[0.1%–0.3%]	[1100]

	Adults and children living with HIV	Adults and children newly infected with HIV	Adult prevalence (%)	Adult and child deaths due to AIDS
LATIN AMERICA				
2007	1.6 million [1.4 million–1.9 million]	100,000 [47,000–220,000]	0.50% [0.4%–0.6%]	58,000 [49,000–91,000]
2001	1.3 million [1.2 million–1.6 million]	130,000 [56,000–220,000]	0.40% [0.3%–0.5%]	51,000 [44,000–100,000]
CARIBBEAN				
2007	230,000 [210,000–270,000]	17,000 [15,000–23,000]	1.00% [0.9%–1.2%]	11,000 [9800–18,000]
2001	190,000 [180,000–250,000]	20,000 [17,000–25,000]	1.00% [0.9%–1.2%]	14,000 [13,000–21,000]
EASTERN EUROPE AND CENTRAL ASIA				
2007	1.6 million [1.2 million–2.1 million]	150,000 [70,000–290,000]	0.90% [0.7%–1.2%]	55,000 [42,000–88,000]
2001	630,000 [490,000–1.1 million]	230,000 [98,000–340,000]	0.40% [0.3%–0.6%]	8000 [5500–14,000]
WESTERN AND CENTRAL EUROPE				
2007	760,000 [600,000–1.1 million]	31,000 [19,000–86,000]	0.30% [0.2%–0.4%]	12,000 [<15,000]
2001	620,000 [500,000–870,000]	32,000 [19,000–76,000]	0.20% [0.1%–0.3%]	10,000 [<15,000]
NORTH AMERICA				
2007	1.3 million [480,000–1.9 million]	46,000 [38,000–68,000]	0.60% [0.5%–0.9%]	21,000 [18,000–31,000]
2001	1.1 million [390,000–1.6 million]	44,000 [40,000–63,000]	0.60% [0.4%–0.8%]	21,000 [18,000–31,000]
TOTAL				
2007	33.2 million [30.6 million–36.1 million]	2.5 million [1.8 million–4.1 million]	0.80% [0.7%–0.9%]	2.1 million [1.9 million–2.4 million]
2001	29.0 million [26.9 million–32.4 million]	3.2 million [2.1 million–4.4 million]	0.80% [0.7%–0.9%]	1.7 million [1.6 million–2.3 million]

An example of the disproportionate prevalence of HIV/AIDS globally. *Source:* UNAIDS/WHO 2007 AIDS epidemic update. *Note:* Sub-Saharan Africa continues to be the region most affected by the AIDS pandemic. More than two out of three (68%) of adults and nearly 90% of children infected with HIV live in this region, and more than three in four (76%) of AIDS deaths in 2007 occurred there, illustrating the unmet need for antiretroviral therapy.

poor settings, is achieved through a practice of "sentinel surveillance." Here, the results of HIV tests among selected groups in a population are used to extrapolate data about infection to the larger collective. These groups typically include pregnant women and, less frequently, people seeking treatment for sexually transmitted infections (STIs). This practice capitalizes on obtaining data from vulnerable people who have some other reason, such as seeking prenatal care, for being caught up in a biomedical apparatus. In industrialized nations, gathering data about disease is easier due to the greater viability of public health infrastructure. For example, many pregnant women in the United States who seek care are tested for HIV; five states mandate such testing, and three states require testing of newborns. This is not the case in many developing nations where whole populations are unable to be captured epidemiologically. Thus, sentinel surveillance, which characterizes the scope of national disease based on the blood of the few, is used to make universal epidemiological claims. These claims are taken up by biopolitical projects that are not necessarily in the best interests of those tested.

Moreover, attempting to classify the scope of the disease has become highly politicized: Is it a pandemic, an epidemic, or endemic? And what are the consequences of each of these definitions? In certain locations, HIV/AIDS has been reclassified from epidemic to endemic. *Epidemic* means a disease spreads at a rate that is greater than would be normally expected—the incidence, or new case rate, exceeds the expectations within a specific geographical region over a specific period of time. One way this could happen is that each diseased individual is able to expose and infect many other individuals. HIV/AIDS has also been labeled a *pandemic*, meaning that the spread of this infectious and contagious disease happens over a large geographic area, such as the globe or a continent. An *endemic* is a disease that is consistently present or native to a particular group of people or a particular geographic area. An infection exists within a given population without exposures from outside the population whereby each individual who has the disease passes it on to exactly one other person. The disease is in a steady state.

Similar to the steady state of HIV/AIDS in certain regions of the world, the disease in the West has been reclassified from an acute and lethal condition with a relatively short disease period and certain death to a chronic disease that can be managed with medication for longer periods of time. Through the use of nucleoside antivirals, non-nucleoside reverse transcriptase inhibitors, protease inhibitors, and maturation inhibitors, people with HIV/AIDS are enrolled in the complicated interactive pharmaceutical process of HAART (highly active anti-retroviral therapy). HAART regulates combinations of these drugs for individual patients with different strains of HIV and different tolerances to drug mixtures.

In the West, drug therapies have led to increased life expectancy for people with AIDS. Popular media have featured stories about people with AIDS now surviving well beyond their expected life span and the growing size of the cohort aged 50 and older living with AIDS.[7] Whereas in previous decades many people with AIDS did not live to see their 50th birthdays, the number of people over 50 surviving HIV increased 77% from 2001 to 2005. And yet, resultant medical complications due to the use of drug therapies with side effects leaves many middle-aged people with AIDS with diminished quality of life that does not reflect their actual chronological age—they appear and function as much older people.

In the developing world, although there are significant regional variations, HIV/AIDS remains an acute and lethal condition. This means that most infected people are dying well before their 50th birthdays. As pointed out by geographer Susan Craddock, "From the 1950s until about 1990, life expectancy for many African countries rose more or less steadily. Botswana achieved a peak of almost 63 in 1990. For South Africa, life expectancies peaked at just over 60 in 1990 and by 2000 were projected to reach 66 without AIDS. Since 1990, these rates have plummeted to 37 and 47 respectively" (2004:3). Death by HIV/AIDS is not only killing people in the developing world, it is also decimating families and communities, creating millions of orphans, severely taxing already compromised public health infrastructures, and contributing to regional instability.

In sum, whereas HIV/AIDS has become largely a manageable chronic disease in industrialized nations, in the developing world the situation is utterly devastating. This huge disparity between wealthy and resource-poor nations in incidence and mortality has created a fertile terrain where new meanings of the disease and responses to it are propagating. We argue that the bimodal distribution of affliction creates two versions of HIV/AIDS and divergent images of AIDS sufferers. In one version, prevalent in the United States, HIV/AIDS is a chronic fact of life, no longer as threatening as it was in the previous century. Visible embodied suffering is relegated to the margins of society, and the AIDS "patient" has become a middle-aged gay man armed with chemical cocktails.

In another version of the disease, pervasive in the death zones of the developing world, HIV/AIDS is this century's black death (Hunter 2003).

Disease is rampant and seeps through bodies and across borders, fostering destabilization—theirs and possibly ours. The quintessential non-Western AIDS "victim" consumed by Westerners is the impoverished, dying African mother on a dirty cot in a makeshift clinic surrounded by her soon to be orphaned children. Instead of inspiring empathy for this mother, such images encourage us to be preoccupied with the potential deviance of her orphans and the unraveling of her country. In the United States, meanings of the disease have become uncoupled from our own vulnerability to HIV/AIDS. Instead, as we show next, our fears are inflamed by images of the afflicted global Other and recoupled with expanding notions of security.

A Matter of National Security, but Whose?

Historian Peter Baldwin (2005) coined the term "geoepidemiology" to describe the intersection of geography, topography, and demography and the ways in which these practices shape a nation's reaction to public health crises. In his investigation of the industrialized world's response to HIV/AIDS, he shows, despite common misconceptions, that the United States was quite interventionist in the early years of the epidemic. For example, new laws were introduced, including the requirement in every state that people with AIDS were to be reported to public health authorities. Civil servants and military personnel were screened, as were prostitutes in some states. Since the 1980s, HIV/AIDS policies in the United States have built on a longstanding quarantinist tradition that has served as a foundation for public health practices for well over a century. Public health scholar Amy Fairchild (2004) describes such policies as having historically set the terms of national belonging via practices of inclusion and exclusion. Both Baldwin and Fairchild highlight as a significant quarantine measure the 1987 bar on immigrants with HIV, first established by the Department of Health and Human Services (DHHS) and later codified in the 1993 Immigration and Nationality Act (INA). If you have HIV, you cannot immigrate to America.

In this policy, we can see the beginnings of nascent security concerns embedded within public health policies. At the same time that humanitarian responses to HIV/AIDS were developing, the U.S. government was also pursuing an aggressive public health strategy of containment. The bar against seropositive immigrants, instituted during a time of considerable public anxiety and fear, was in theory designed to seal our borders against those carrying deadly HIV. Never mind that the afflicted were already inside the nation infecting each other. Most major medical and public health organizations opposed the bar, and it had detrimental effects on the nation's efforts to fight HIV/AIDS globally. For example, in protest against the policy, organizers of the International AIDS Conference have refused to hold an annual meeting in the United States since 1990. Yet the bar remains in effect at the time of this writing in 2008, and has had the most serious impact on immigrants wanting to relocate permanently to the United States.

One of the most disturbing consequences of the policy can be seen in the treatment of Haitian refugees who fled their country in 1991 after the overthrow in a military coup of popularly elected Jean-Bertrand Aristide. Some 40,000 Haitians attempted to enter the United States seeking asylum status, many arriving by sea where they were interdicted and taken to Guantanamo Bay. Approximately 10,500 were granted admission to the United States, and 25,000 were sent back immediately. Those granted asylum were tested for HIV, and approximately 200 people tested positive; some were returned to Haiti. A group of 158 people (143 seropositive adults, two seronegative adults, and 13 untested children) were housed for almost two years at a detention facility at Guantanamo Bay. According to anthropologist Paul Farmer, the conditions at this facility were brutal. Farmer quotes one woman who initiated a hunger strike among detainees. She told him, "Bees were stinging children, and there were flies everywhere: whenever you tried to eat something, flies would fly into your mouth" (2005:64). Others described razor wire, solitary confinement, and a humiliating lack of privacy. The facilities were also clearly insufficient for adequate treatment of people suffering from HIV/AIDS. Judge Sterling Johnson of the U.S. District Court determined that the facility was "nothing more than an HIV prison camp" (61). Following a protracted legal battle, all 158 detained Haitians were eventually allowed into the country.

Sixteen years later, Human Rights Watch (HRW) reported that seropositive detainees at immigrant detention centers around the country were not receiving effective care for HIV/AIDS. In a 71-page report entitled "Chronic Indifference,"[8] the HRW criticized U.S. Immigration and Customs Enforcement (ICE, formerly INS) for its negative treatment of detainees. A division of the Department of Homeland Security, ICE is responsible for, among other things, managing a network of immigration detention centers mostly located in border states. These centers engage in "detention and removal operations," a euphemistic term referring, presumably, to the handling of illegal, sick, or otherwise unfit human bodies that according to regulations do not belong in the United States. While the concerns raised in the HRW report are not insignificant, their details pale in comparison with stories about the Haitian experience at Guantanamo Bay. Yet what connects these two episodes in the nation's immigration history is the intersection of public health concerns with the apparatuses of national security. Foreign bodies are quarantined in the service of a coherent national identity and the safety of the majority.

Cut to April 2000, when President Bill Clinton's administration initiated a major policy shift, for the first time marking HIV/AIDS as an issue of national and global security. A White House spokesperson described AIDS as "more than a legitimate ongoing health threat," noting that the disease "also has the potential to destabilize governments such as African or Asian nations, which makes it an international security issue." The new policy came on the heels of a report from the National Intelligence Council, stating that AIDS "will endanger U.S. citizens at home and abroad, threaten U.S. armed forces deployed overseas, and exacerbate social and political instability in key countries and regions in which the United States has significant interests."[9] The security designation brought an infectious disease under the purview of the National Security Council (NSC) for the first time in the agency's history, expanded federal research on HIV/AIDS to include the Department of Defense, and inaugurated a concentrated effort of federal agencies in the process of rearticulating the scale and scope of the disease.[10]

At the time of the Clinton policy shift, some Republicans evidenced a fairly predictable reaction, suggesting that AIDS was not in fact a security threat. Senate Majority Leader Trent Lott, R-Mississippi, claimed on Fox News, "This is just the president trying to make an appeal to certain groups." Lott said he did not consider AIDS to be a threat to national security—"not our national security."[11] Yet in 2002, just two years into the new Republican Bush administration, Secretary of State Colin Powell stated, "HIV doesn't just destroy immune systems; it also undermines the social, economic, and political systems that underpin entire nations and regions."[12] In 2003, CIA chief George Tenet argued, "The national security dimension of the virus is plain: it can undermine economic growth, exacerbate social tensions, diminish military preparedness, create huge social welfare costs, and further weaken already beleaguered states. And the virus respects no border."[13] So what happened in the period between the Clinton policy designation and Republican disavowal of it, and the subsequent embrace of the national security framework by the Bush administration? In a word, or rather a number, it was 9/11 and the ensuing rapid, commercialized development and deployment of a variety of security strategies, tactics, and products. Similar to the Clinton administration, then, the Bush administration in the post-9/11 era pursued an agenda of containment via security discourses.

The framework of HIV/AIDS as a security problem has proliferated exponentially and been disseminated widely, in both partisan and nonpartisan ways. In a 2002 issue of *Foreign Affairs,* neoconservative political economist Nicholas Eberstadt drew attention to the Eurasian pandemic affecting Russia, India, and China. Drawing on unsubstantiated "intelligence estimates" about the magnitude of HIV/AIDS in these countries, he positions them as emerging threats to global security and "a looming catastrophe." He analyzes the economic consequences of the disease both within these nations in terms of effects on human capital and beyond their borders as they attempt to enter transnational economies. Eberstadt warns, based on access to epidemiological data collected by the U.S. intelligence community,[14] that "these crises are only beginning." He forecasts three potential scenarios, termed mild, intermediate, and severe, in which he models how the pandemic might play out in each country given a certain set of circumstances. He writes, "Population growth, labor supply, and savings rates will all be hurt—indeed the more comprehensive the framework employed, the more negative the conclusions seem to be" (2002:38). In short, Eberstadt paints a grim picture of Russian, Chinese, and Indian futures; poised to enter the global economy, they may be undermined by their own histories and the debilitating impact of HIV/AIDS.[15]

Policy analyst Greg Behrman chronicles the failure of the United States to respond appropriately to the pandemic, particularly in Africa. In a critical analysis of political intrigue and policy debacles, he highlights the security aspects of HIV/AIDS in unnecessarily dramatic language. He states, "The disease will have grave consequences for U.S. and global security. 9/11 demonstrated the danger of a failed state. In Africa, the world confronts nothing less than the potentiality of a failed continent. . . . The debris, without drastic action, will be the fodder upon which terrorists and transnational criminal elements will find refuge and sustenance" (2004:xii). He draws attention to the behind-the-scenes maneuvering of people in the Clinton administration who were attempting to reposition the disease. One such person was Senior Advisor for International Health of the NSC Dr. Ken Bernard, who pragmatically recognized that, according to Behrman, "if an issue was deemed of national security import, it attracted funds with the velocity that no other area of U.S. government can match" (229). In other words, HIV/AIDS had to be repackaged to warrant the type of funding and attention that was being directed to other perceived threats.

In 2005, the Council on Foreign Relations released a report authored by journalist Laurie Garrett titled *HIV and National Security: Where Are the Links?* The 2005 report has been widely cited as evidence of a link between HIV/AIDS and national security threats, yet Garrett's approach is actually more complex and critical. She writes, "Many governments view domestic instability as a primary national security threat. Individuals infected with HIV and their advocates have been labeled threats to the state. This is clearly a repressive approach to the epidemic that is rooted in two falsehoods: first, that HIV patients and their advocates seek to overthrow the state; and second, that outbreaks of the virus can be limited through repression of those individuals it infects. There is no evidence that people infected with the virus that causes AIDS, in any country, posed a direct threat to state security" (10).Yet she does raise the possibility that destabilization in HIV-ravaged nations could foster anti-Western sentiment.

In 2006, epidemiologist and global health expert Harley Feldbaum and colleagues issued a call to public health experts to further explore and advance linkages between public health and security discourses, primarily as a means of garnering international support and funding. They write, "What is clear is that arguments linking HIV/AIDS to national security have succeeded in elevating the disease to the highest levels of international politics, resulting in greater political commitment and funding" (777). As evidence of links to security, they cite HIV/AIDS rates among soldiers in many countries, destabilized nuclear states, the breakdown of families and orphaning of children, and famine brought about by the death of agricultural workers. Yet Feldbaum et al. are cautious about certain aspects of security discourse. They admit that the objectives of national security and those of public health may diverge: "Global health works to improve the health of all people within and across states, while the national security field works to protect the people, property, and interests of only one state" (774). In the end, they seem to want to reclaim HIV/AIDS as a public health concern but one that is understood through a security lens offering greater political clout.

In the post-9/11 policymaking arena, then, claims to security needs flourished and were seen as legitimate. Indeed, there appeared to be the rise of a new security imperative. While personnel in the Clinton administration realized that connecting humanitarian and public health concerns of HIV/AIDS to the security apparatus would bring more attention and funds, this pragmatic approach in the Bush years was amplified by the emergence of an entire biosecurity apparatus, which further rendered intelligible the national security aspects of HIV/AIDS. In other words, Clinton and his staff, operating in typical pragmatic fashion, initiated a policy shift that, after the attacks of 9/11, became inexorably and rapidly caught up with corporate, military, and administrative maneuverings of Bush's War on Terror. Indeed, so successfully was HIV/AIDS repositioned as a component of the national security strategy, that it now seems inconceivable that it was ever understood as anything else.

Not only are policymakers and scholars discussing and promulgating HIV/AIDS as a security issue, but also resources are being targeted to the disease and its consequences, including "failed states." We argue that these allocations both reflect and contribute to ongoing citizen fear and anxiety. The American "soccer mom" is no longer afraid that she or her children will become infected with HIV; she is now (presumed to be) afraid of being attacked by terrorists. The condom becomes a throwback to a past American era, while the noncombat, retail version of the Humvee embodies a new suburban prophylactic promising to secure the bodies and lives that are deemed to matter most. (At least, this was the case until escalating fuel prices prompted middle-class and upper-class consumption of the more economically and ecologically savvy Prius.)

Capitalizing on post-9/11 anxiety and emergence of new constituencies, in 2003 the Bush administration allocated $15 billion (to be spread over five years) to fight AIDS, citing national security concerns. The President's Emergency Plan for AIDS Relief—PEPFAR—was a multibillion-dollar aid package for international public health initiatives to combat the spread of HIV/AIDS and treat those infected. In his position as secretary of State, Colin Powell repeatedly registered his (gendered) concern that deaths due to AIDS "threatened to wipe out the entire child-bearing population—a condition that could create instability, and a climate ripe for terrorism."[16] Powell also recalled a conversation with Tommy G. Thompson, the DHHS secretary: "I said, 'Tommy, this is not just a health matter, this is a national security matter.'" Critics alleged that PEPFAR was a politically expedient action to create global good will toward the United States, then on the brink of invading Iraq. Michael J. Gerson, Bush's speechwriter, attempted to dissuade the public of this thought but admitted that PEPFAR was part of the "spreading democracy" agenda of the Bush White House.

PEPFAR, in partnership with host countries, seeks to establish local programs that support the uptake of key prevention behaviors. The "ABC" approach encouraged by PEPFAR claims to tailor behavioral messages to local epidemic contexts but in reality offers a one-size-fits-all universal approach reflective of American conservative values: "A" behaviors include abstinence, or delay of sexual debut for young people; "B" includes faithfulness to one partner or reducing the number of sexual partners; "C" emphasizes correct and consistent condom use, where appropriate.[17] Health policy journalist Nellie Bristol reported, "At least 50 percent of PEPFAR funds used for the prevention of sexual transmission of AIDS shall be dedicated to abstinence and fidelity . . . and the correct and consistent use of condoms, consistent with other provisions of the law and the epidemiology of HIV infection in a given country" (2007). Funneling PEPFAR funds through American religious organizations in other countries and attaching the "ABC" requirement represents a form of neocolonialism by which the United States proselytizes about appropriate values and attempts to socially engineer other cultures and nations via restrictions on sexual behavior. This is consistent with the Bush administration's approach, domestically and internationally, to other reproductive and sexual issues including family planning and suggests that, while the security framework may be nonpartisan, the ways in which it is implemented are deeply partisan.

The Bush administration also initiated Project BioShield, which was signed into law in July 2004. A key feature of the emerging biosecurity apparatus, BioShield was designed to bolster the biodefense industry. Funded with $5.6 billion, the fund enables research and manufacture of technologies and devices for countering chemical, biological, radiological, or nuclear (CBRN) attacks. Corporations apply for R&D funding from the U.S. government to invent and produce biodefense countermeasures and are also contracted to implement government projects. One such device was Aethlon Medical's Hemopurifier, a hollow-fiber hemodialysis cartridge that removes pathogenic viruses, bacteria, and toxins from an individual's blood. Aethlon Medical's Internet site describes the technology as a positive solution to the problem of "viral conditions that are either resistant or evolve resistance to drug and vaccine therapies." (Gostin and Fidler 2006/2007:452).

Legal scholars Lawrence Gostin and David Fidler define biosecurity "as a society's collective responsibility to safeguard the population from dangers presented by pathogenic microbes—whether naturally occurring or intentionally released" (2006/2007:438). Both PEPFAR and BioShield are biosecurity strategies of a government that is determined to protect and reinforce our borders, corporeal and geopolitical. These programs provide funding for enhancement of public health infrastructure but in limited ways, at the same time that other aspects of public health are severely defunded (Garrett 2000). Responses to biothreats are emphasized over basic preventive care or access to HAART. In the new security framing, the bodies of the afflicted become both more and less visible in paradoxical ways; in the United States, domestic sufferers are largely undetectable, whereas in the developing world, decrepit and dying bodies are subsumed within the geopolitical object of the "failed state." Foreign bodies in the aggregate become perilous and threatening, challenging our already precarious and manipulated sense of security in the post-9/11 era. Their tragic affliction inspires biodefense rather than empathy and care. Thus, in reframing HIV/AIDS as a security issue, policymakers have laid essential groundwork for it to recede as a global health or humanitarian concern and for the bodies of the infected to disappear from public view.

Superinfectors and Other Terrorists

Cultural geographer Bruce Braun has explored the paradigm shift from population risks to a new somatic self that is concerned with the work of biosecurity at the molecular level. His argument foregrounds the precarious body facing imminent catastrophe; the more vulnerable a person with respect to social stratification—for example, undocumented workers, the poor, the uninsured—the more precarious she is and the more likely to be exposed to risk. He writes, "what we find in the medical and political discourse of 'emerging infectious diseases' is a body that is radically open to the world, thrown into the efflux of an inherently mutable molecular life where reassortment is not what we control, but what we fear. This post-genomic world is not understood in terms of one's genetic inheritance—nor is it primarily about 'care of the self' or 'genetic citizenship'—it is instead understood in terms of a global economy of circulation and exchange that at once precedes and transcends the individual body" (2007:18). What we have shown regarding the growth of a discourse focused on gaining health through security is that, as geographer Braun frames it, "public health remains a geopolitical exercise concerned with the sanctity of borders, dangerous migrations and foreign risks" (22).

One important consequence of the shift to security discourses and practices is that some bodies afflicted with HIV/AIDS, and other bodies positioned as biothreats, become hypervisible at the same time that the individual bodies of victims in the developing world are disappearing, actually and symbolically. We argue that with the rise of the security apparatus and a declining focus on domestic HIV/AIDS, the cases that do capture our attention (competing with reality TV and infotainment) are the outliers, the scary bogeymen who loom large as monsters whose stealth and unpredictability keep us in a state of perpetual vigilance. For example, terrorists have become the new virus piercing the national body and wreaking havoc on our homeostasis and myth of social order. Since the 1980s, flights of hysteria and the rhetoric of risk and infection have been deployed regarding gay men especially, drawing on homophobia and conservative beliefs about normative human sexuality. Domestically, media attention about HIV/AIDS makes visible and highlights what we call dramatic bodies that evoke an imagined seedy, underground community of deviants reminiscent of the early days of AIDS.

In 2001, stories about African American men on the "down low" (DL)—that is, men who have sex with other men but conceal this behavior from female partners—hit the airwaves. J. L. King, author of *Living on the Down Low*, appeared on a segment of *The Oprah Winfrey Show*. A 2007 human-interest story in *Essence* magazine titled "Living on the Down Low" profiled a man named Keith Farmer, who stated, "The hardest part about being on the DL was the creeping and deceiving of women. It made me feel so guilty" (Brown 2007:122). Over the past several years, the DL has been a figure trotted out on TV talk shows and in print media in profitable ways without critically investigating actual sexual practices. According to a team of epidemiologists led by Chandra Ford, the DL is used incorrectly to discuss a made-up phenomenon of black sexuality, yet there is no evidence to support that the "DL" is either new behavior or is limited to African American men. It also reflects "social constructions of black sexuality as generally excessive, deviant, diseased, and predatory" (Ford et al. 2007:209). Journalist Keith Boykin (2005), who is also gay, argues similarly that the DL phenomenon misrepresents African American male sexuality and links to HIV/AIDS.

Criminalizing those with infectious diseases is not a new practice. As articulated by Kai Wright, the editor of BlackAIDS.org, criminalization of those infected with HIV is illustrated through a state-sanctioned policing of their behavior as a method of disease control. Nondisclosure of HIV-

Cover of book. Keith Boykin, *Beyond the Down Low: Sex, Lies, and Denial in Black America* (Perseus), 2005.

positive status to sexual partners is a potential crime that can be used in any state to prosecute, whether or not transmission occurred. A racialized urban legend exists of the seropositive man of color as a "superinfector" lying about his HIV status and knowingly and purposefully infecting as many people as possible. As Wright notes, "'super infector'—[was] a riff on the then-popular 'super predator' caricature for young, urban Black men" (2005).

In addition to the superinfector, other dramatic bodies include those labeled "barebackers" and "bug chasers." In the late 1990s, the term "barebacking" rose to prominence to describe the practice of men intentionally having insertive anal intercourse with other men without condoms (Halkitis et al. 2006). Both the gay and mainstream media picked up on this phenomenon, reporting with varying degrees of outrage and dismay unsubstantiated claims of the prevalence of such practices framed as "epidemic." Bug chasers are a subculture of men who have sex with men in an attempt to voluntarily contract HIV (Moskowitz and Roloff 2007). As historian David Halperin (2007) points out, stigmatization of unsafe sexual practices is a way of re-pathologizing homosexuality—and those who engage in such practices (Halperin 2007). We are not questioning the existence of these men but, rather, are claiming that the terms used to describe them are derogatory and loaded. They may be useful in epidemiological and subcultural worlds but may not be empirically significant. How does such a presumably small group of "deviant" men become visible and represented—indeed, overexposed—in a way that diverts attention from other issues, bodies, and realities?

From the "illicit" behavior of these three groups of bogeymen—the DL, superinfectors, and bug chasers—emerges a new category of virus, superbugs. A *New York Times* article in January 2008 linked an emergent strain of the staph bacteria to the behavior of gay men. MRSA (methicillin-resistant *Staphylococcus aureus*) is a multidrug-resistant bacteria that gay men were "many times more likely than others" to acquire.[18] Inevitably, conservative groups such as Concerned Women for America used the report as ammunition in their attempts to patrol certain bodies and behaviors: "Homosexual conduct is a breeding ground for deadly disease."[19] While the nation's security apparatus is aimed at failed states, within our own backyard the notion of deviant bodies continues to fuel partisan cultural anxiety about normative sexuality. Both sets of practices, domestic and global, are about policing bodies and mitigating risk in the service of broader political agendas.

From Red Ribbons to (RED)

In 1991, people began sporting red ribbons on their lapels. The Red Ribbon Project was officially launched at the Tony Awards in New York City. Its purpose was to indicate solidarity with and among people with AIDS and to foster AIDS awareness. The ribbons also soon came to be used to memorialize the dead. These tiny flashes of fabric represented the suffering of people afflicted with HIV/AIDS. In 1990s San Francisco, we wore the ribbons and recall seeing them on many of our friends and colleagues. In 2008, red ribbons are hard to spot (unlike the yellow LiveStrong bracelets, which we discuss in chapter 7). In lieu of red ribbons, we now have the (RED) campaign, an effort in philanthropic consumption (Talley and Casper, forthcoming) in which buying "red" results in contributions by participating corporations (e.g., Gap, Motorola, Apple, and many others) to the Global Fund to Fight AIDS, Tuberculosis, and Malaria. These two humanitarian projects, centered on the color and symbolism of blood and emergencies, both focus on suffering bodies. But whereas one proliferated in the pre-9/11 era when public health models were dominant, the other has been profoundly shaped by its intersection with security discourses.

One question driving our research was this: If public health models have receded, who in the contemporary moment frames HIV/AIDS in terms of humanitarianism? Where are the spaces and places in which the disease is considered in "old-fashioned" terms as an embodied health risk? With the media, policymakers, and scholars promoting the security framework, the job of publicizing and advocating for a health and human rights approach has apparently fallen to celebrities and activists. A chief donor to and spokesperson for the Global Fund is rock star Bono of U2 fame, who in 2007 guest edited a special issue of *Vanity Fair* on Africa. While violence and structural adjustment are highlighted in the magazine alongside stories of suffering and loss and glossy advertisements for products, we are most struck by the pervasive focus on "African AIDS," which in our view reflects the ways in which a security approach has channeled the diverse spectrum of global HIV/AIDS into one disastrous region of strategic interest to the United States.

Of the 20 "historic" cover photos taken by Annie Leibovitz, more than one-half of the subjects reference AIDS, and Bono's guest editorial reads, "Next year, more than 10 million children's lives will be lost unnecessarily to extreme poverty, and you'll hear very little about it. Nearly half will be on the continent of Africa, where H.I.V./AIDS is killing teachers faster than you can train them and where you can witness entire villages in which children are the parents" (2007:32). Yet in exposing HIV/AIDS as a pubic health and humanitarian concern, a few chosen celebrities also deploy the bodies of the diseased and ravaged to expose their own marketable bodies and careers to an adoring public, suggesting that participation in philanthropic efforts may have as much to do with perpetuation of celebrity status as with lasting social change. Celebrities and issues may become more visible, but the bodies of the sufferers fade into the background.

Immediately after the attacks on the World Trade Center in 2001, President Bush attempted to soothe a worried public. In an appeal to traumatized citizens, he suggested that we respond to the terrorist attacks by supporting the American economy; this phrase was repeated in multiple media outlets and interpreted as urging us to "go shopping." We were also asked to "be patient with security measures." The conflation of national security with political economy following 9/11 suggests to us that spaces for democratic practices, for meanings of life not connected to security, and for public resistance to the emerging apparatus including the War on Terror had already begun to shrink. They have continued to do so, and the future is uncertain. We contend that when concerns such as HIV/AIDS—a public health crisis of epic magnitude—become the jurisdiction of national security infrastructures intent on surveillance and border patrol, they may cease to be recognized as freedom or rights projects.

As we show in chapter 5, security discourses also seep into other biopolitical practices, such as human biomonitoring, which is used to assess the impact on organic bodies of living in an increasingly toxic world. Whereas this chapter explored meanings and consequences of viral invaders in human bloodstreams, next we examine what happens when contaminated fluids are extracted from the body for purposes of scientific evaluation.

5

Fluid Matters

Human Biomonitoring as Gendered Surveillance

May the world's feast be made safe for women and children. May mothers' milk run clean again. May denial give way to courageous action.

Sandra Steingraber, *Having Faith* (2001)

By any reckoning, planet Earth is in dire straits. Some might say that humans have been imperiled ever since Eve snapped the forbidden fruit off the tree, bit into it, and handed it to Adam. Original sin is certainly one popular theory for why we as a species are chronically beset by tragedy and suffering. Another version of the story, one prevalent in an ever-expanding literature on environmental toxins, points the finger at synthetics that are poisoning our habitat and causing trouble in paradise. The real problem, scholars argue, is not that Eve flirted with a serpent but that humans have for too long engaged in intercourse with another kind of devil: industrial chemicals. From hazardous waste to global warming, the abacus by which we increasingly make sense of our precarious place on the planet is ecological. We seek to measure, calculate, and evaluate the degree of risk posed by pollutants that, for over a century, have been seeping into our air, soil, water, and bodies, transforming our exterior and interior landscapes. We have become environmental subjects, progressively more definable by external risks to our bodies and a host of surveillance technologies developed to assess these threats.

One such technology is biomonitoring, the measurement of trace chemical compounds in human beings using molecular biological techniques and measurement of exposure using analytic chemistry. Biomonitoring is an extractive technology; that is, liquids and tissues must be removed from human bodies in order to be manipulated and assessed. Thus, it is not whole bodies that are being measured but leaky and diaphanous bits and parts, such as breast milk and blood; these stand in for and are made to represent whole persons—indeed, entire communities and the species writ large. Biomonitoring enables the partition of human bodies, offsite treatment of excised and discharged material in a laboratory, calculation of its properties, and subsequent symbolic concealment of the body's constitutive materials within abstract numeric codes. Whole bodies, then, through a series of iterative processes are translated into data representing individual and species risk; bodies themselves are reconfigured as toxic waste dumps. Just as actual toxic waste dumps are hidden from most people's everyday view, so does the private, remote processing of the body's toxic debris remain hidden from the public. It is merely data and not the actual "human landfill" that we see.

Although not new, biomonitoring has been taken up in fresh ways as a technical strategy for assessing human embeddedness in compromised ecologies. It has also been the subject of serious debate. For example, in 2003 California State Senator Deborah Ortiz introduced S.B. 689, a bill to establish the Breast Milk Biomonitoring Pilot Program. Cosponsored by health-advocacy organizations Breast Cancer Fund and Commonweal, the bill would have provided funding to develop a statewide program to measure toxins in people's bodies, or what is known as their "body burden." Using the tools of biomonitoring, Ortiz and her supporters argued that the resulting data about exposure would be useful to scientists, medical professionals, public health officials, policymakers, and community members by showing which harmful chemicals, industrial compounds, and pollutants had taken up residence in human bodies. Despite significant community support, the Senate Committee on Appropriations refused to hear the proposed legislation, and it never came to a vote.

In 2004, Senator Ortiz again introduced a biomonitoring bill, S.B. 1168, this time bolstered by efforts of the California Body Burden Campaign (CBBC), a network of environmental, health, faith, labor, medical, and women's organizations.[1] S.B. 1168 was poised to create the first permanent, statewide, community-based biomonitoring program in the nation. Building on experience gleaned during the unsuccessful 2003 legislative session, the bill's supporters launched a public education campaign, conducted research on attitudes toward environmental illness,[2] and expanded the CBBC's website to include a new slogan, "It's time to start looking at what's in our bodies." In spite of vocal opposition from industry, the Senate passed S.B. 1168 and the State Assembly Environmental Safety and Toxic Materials Committee also approved the bill. However, it narrowly missed passage in the Assembly Health Committee, by just one vote.

The following year, a powerful coalition of industry groups opposed the bill's next version, S.B. 600, making their case in an open letter to Senator Ortiz.[3] Their key concerns were that the proposed legislation lacked a framework for interpreting biomonitoring results, that there was no science-based criteria for developing the program, that the scientific advisory panel's role was unclear, and that the legislation posited a cause and effect relationship between detection of a chemical and adverse health outcomes. Although adopting an oppositional stance, the coalition indicated support for biomonitoring if it could be done under appropriate conditions; industry representatives would "welcome the opportunity to enter into a constructive dialogue."[4] Their resistance to the bill illustrated longstanding claims that biomonitoring legislation was politically motivated—that is, politicians were succumbing to constituents' activist concerns—rather than truly scientific.

In 2006, Governor Arnold Schwarzenegger at last signed biomonitoring legislation, S.B. 1379, making California the first state to embark on such a comprehensive program. By that time, the original bill had been thoroughly transformed through the input of legislators and stakeholders, including such unlikely couplings as the Breast Cancer Fund and the chemical industry. (For insight into similar alliances in breast cancer worlds, see Klawiter 2008.) The American Chemistry Council reported that the industrial coalition removed its opposition to the bill "based on the adoption of amendments addressing longstanding concerns about the need for science-based policies."[5] Among the changes in the successful legislation were a clearer statement of the research objectives of the program, ensuring a statistically valid and representative sample of participants, providing a formal structure for research design, specifying ethical standards for communicating data, and establishing an advisory panel of qualified experts to serve as peer reviewers. Indicating the melding of "scientific" and "economic" concerns and the complicity of government with industry, Department of Toxic Substances Control director Maureen Gorsen remarked, "Biomonitoring is an important tool to help California businesses develop cleaner technologies, keeping our state in the forefront of innovation."[6]

Shortly after the bill was approved, the California Department of Health Care Services (CDHCS) and the California Department of Public Health (CDPH) together helped to develop the California Biomonitoring Plan (CBP), but it was never administered. CDHCS and CDPH made claims linking their CBP program and its technologies to broader public health initiatives: "Biomonitoring data supports public health by establishing trends in chemical exposures, validating modeling and survey methods, supporting epidemiological studies, identifying highly exposed communities, addressing the data gaps between chemical exposures and specific health outcomes, informing health responses to unanticipated emergency exposures, assessing the effectiveness of current regulations, and helping to set priorities for reform." The program's Internet site emphasized that the bill requires *public* access to information, acknowledging the insistence of various stakeholders that information be widely available.[7]

As illustrated in our brief summary of the clamor surrounding biomonitoring in California, this technology is by no means neutral. Proponents and opponents adhere their divergent perspectives to it like so many partisan stickers to the bumper of a minivan or pickup truck. Analyzing these debates reveals multiple meanings of biomonitoring, to be sure. But such an analysis also makes visible extant meanings about human bodies and body parts, especially those that are significantly gendered. It is no accident that the contention related to the 2003 legislation, which highlighted the importance of breast milk, made for very strange bedfellows and temporarily pitted breast cancer activists seeking better health and a cure against breastfeeding advocates seeking better nutrition for babies.

We show in this chapter that the fluids of men and women—semen and breast milk, in particular—represent hallmarks of gender. When separated from their source, each gets cut off from the bodies of the people who produce the fluid in the first place. These bodies disappear, and just the liquid remains. Yet the essence of these bodies leaves its mark on how the fluid is understood—including its value, its use, and its very properties. Women's fluid is thought to save others, in a vivid representation of Mother Earth and mothers of the earth at work, whereas male fluid is a renegade, an outlaw, or an impregnator. Given what we know about our gendered expectations and practices, this is not surprising. But it is important to understand how this happens and to question it.

In what follows, we explore these "fluid matters" first through an overview of the technique of biomonitoring. Then we offer a critical and skeptical account of breast milk as the "gold standard" of biomonitoring contrasted to the surprisingly infrequent use of another promising substance: semen. We do not claim here to present a definitive analysis of the science and contemporary uses of biomonitoring; other scholars are ably pursuing this topic (e.g., Washburn, forthcoming; Altman 2008). We do, however, turn our lens onto this gendered technology as a way of making sense of a biopolitical field in which actual human bodies are among the missing at the same time that liquids, tissues, and molecules are made visible and circulated through a new surveillance regime of knowing and representing.

Rise of Human Biomonitoring

Beginning in the 20th century, occupational physicians and industrial hygienists used biomonitoring techniques to assess workers for exposure to hazardous substances. In the 1970s, biomonitoring was a crucial component of efforts to prevent childhood lead poisoning. After learning that excessive levels of lead in blood could result in severe cognitive deficits, policymakers made a concerted effort to remove lead from gasoline, paint, and other products. The lead-free movement was partially successful in the United States; for example, fuel sold at American gas stations contains no lead. Yet children in many urban environments may still be exposed to lead in old paint, leaded gasoline is used throughout developing nations, and lead persists in soil—all suggesting that the toxin continues to be a public health problem for some communities (Schettler et al. 1999). Now, decades after the lead-free movement, according to *Science*, "biomonitoring is hot . . . such studies can generate headlines and political leverage" (Stokstad 2004:1892). Public health expert Lauren DiSano asserts, "The time has come to expand beyond lead protection programs and to begin monitoring for new and emerging exposure hazards and to identify 'hot spots' where public health and public safety may be at risk" (2006:32)

The term "biomonitoring" refers to "the analytical measurement of biomarkers in specified units of tissues or body products. . . . These biomarkers are any substances, structures, or processes so measured that indicate an exposure or susceptibility or that predict the incidence or outcome of disease" (Albertini et al. 2006:1755). Blood, urine, breast milk, semen, and other tissues and fluids are analyzed for the presence of "any alteration in cells or biochemical processes that can be measured in a biological system or sample" (Becker et al. 2003:2). As a scientific enterprise that combines epidemiology, demography, and environmental science, biomonitoring measures and compares biomarkers of exposure, which detect the original chemical, and biomarkers of effect, which assess a biological response. The two different biomarkers mean that, in some cases, the chemical itself may be the measurement; in others, the marker may be some change in the body caused by the action of the chemical. Exposures are complex and may be synergistic; that is, chemicals together may produce adverse effects that they might not produce alone. Biomonitoring can measure naturally occurring chemicals such as arsenic and lead, as well as synthetic or man-made chemicals such as pesticides and polybrominated diphenyl ethers (PBDEs). Given that there are some 85,000 synthetic chemicals on the planet—most of which have not been tested for toxic effects—biomonitoring has emerged as an important strategy and topic, not just for scientists but also within a growing transnational environmental justice movement.

Environmental health scientists Dennis Paustenbach and David Galbraith state, "each of us has ingested, inhaled, or absorbed a variety of these chemicals . . . As a result of well-conducted biomonitoring studies, we can obtain a picture of the amount of chemical or agent actually absorbed into the human body" (2006:1143). They suggest that advances in analytical chemistry, including the ability to detect extraordinarily small concentrations of chemicals in a substance, as well as new toxicological and epidemiological methods, have made possible enhanced and expanded biomonitoring. Toxicologist Richard Albertini and his colleagues note, "Public and private demands for biomonitoring data are on the increase" (2006:1755). As evidence, they point to U.S. government programs such as the National Health and Nutrition Examination Survey (NHANES) conducted by the CDC, which uses data collected to produce the National Report on Human Exposure to Environmental Chemicals. The Environmental Protection Agency (EPA) conducts pilot studies as part of the National Human Exposure Assessment Survey (NHEXAS), and the National Institute of Environmental Health Sciences (NIEHS) sponsors research on environmental exposures in several Centers for Children's Environmental Health and Disease Prevention Research around the country.

As stated earlier, biomonitoring produces a measurement referred to as "body burden," or the total amount of chemicals that have bioaccumulated in the human body and can be assessed at a given point in time. There are numerous ways toxins enter the body, including ingestion, inhalation, and absorption; pregnant women can pass substances to their fetuses through the placenta and umbilical cord, and nursing mothers can pass toxins through breast milk. Some chemicals stay in the body for only a short time; arsenic, for example, usually leaves the body within 72 hours of exposure. However, continuous exposure to many chemicals can create a "persistent" body burden by lodging in fatty tissue, semen, muscle, bone, brain matter, and other organs. DDT (dichlor-diphenyl trichlor), a chlorinated pesticide, can remain in the body for up to 50 years. Biomonitoring can thus reveal an individual's unique "chemical load"—the result of a lifetime of bioaccumulation—highlighting the kinds of chemicals a person has been exposed to from before birth to the point at which the study is conducted.

Media coverage of biomonitoring reveals fascination and repulsion with the notion of a body burden. In 2003, Mount Sinai School of Medicine, Environmental Working Group (EWG), and Commonweal conducted a heavily reported study in nine adults, measuring 167 chemicals in the subjects' blood and urine at a cost of $5,000 per person. The study, called "Body Burden: The Pollution in People," found an average of 91 compounds in each person.[8] *Wired* magazine documented the study: "Davis Baltz shops for organic food and otherwise tries to live as healthy as he can. So he was shocked to learn that the pollutants collecting inside his body sounded much like a Superfund cleanup site: pesticides, flame retardants and other nasty, man-made chemicals turned up in a recent test" (Associated Press 2003) Covering the same investigation, the *Ottawa Citizen* reported, "Davis Baltz is a toxic waste site, according to a 2003 study that unearthed 15 dioxins and furans, 41 PCBs, four organochlorine pesticides, 33 volatile and semi-volatile organic compounds, lead, mercury and phthalates. Problem is Davis Baltz is not a place, he is a person" (Allan 2006).

Just as individuals have a body burden, so does the planet; indeed, this relationship between humans and their habitat is precisely what propels environmental assessment technologies. Thus, biomonitoring is of increasing transnational and global relevance—unsurprising, given that toxins, like people, can and do migrate across geopolitical borders. In 2005, the U.K. division of the World Wildlife Fund offered a media-friendly "biomonitoring tour" to 155 people, including members of Parliament and journalists. Results showed over 70 chemicals in these people's bodies, with an average of 27 chemicals in each person tested. The World Health Organization (WHO) has already funded several scientific studies of biomonitoring, and in 2003 the Commission of European Communities published "A European Environment and Health Strategy," which recommended widespread and consistent biomonitoring around the world. It remains to be seen whether biomonitoring will be as prevalent in developing nations as it is in the developed world, given other pressing concerns in the Global South, such as poverty, hunger, and HIV/AIDS. Yet there is considerable interest in using the technique to explain the high levels of contamination in populations living in even remote areas, such as the Inuit in Greenland and indigenous groups in regions of the Arctic (Schafer 2006). We must ask: Whom will this knowledge benefit?

While an expanding body of scientific studies attest to the usefulness of biomonitoring and investigations of body burden, caution and controversy remain among members of scientific communities, industry, and the public. As critics argue, the data is meaningless without a context for interpreting it; we need to know not only that there are chemicals in our bodies but also what quantity of these chemicals pose a health risk, for what period of time, and for whom. Journalist Erik Stokstad reports in *Science*, "although biomonitoring can provide reams of statistics about chemicals people are exposed to, it can't necessarily indicate whether such exposures are likely to make them sick. . . . What's becoming even more obvious, researchers say, is a growing data gap: Although testing for a chemical can take just a few days, discerning its impact on health takes years" (2004:1892). Measurement of a substance in a specific sample does not, in fact, provide data about the source, timing, frequency, magnitude, or duration of exposure. The chasm between biomonitoring results and "proof" of their impact on human health creates a situation conducive to debate. Critics of biomonitoring, some of whom refer to it as "junk science" (Bast 2004), point to interpretive issues as indications that the measurements are political rather than scientific and that organizations that support biomonitoring are generally "anti-chemical" and "anti-business." Note that this is the same criticism targeted at "popular epidemiology" and other citizen-driven scientific endeavors (Brown and Mikkelson 1997).

One way these perceived limitations of the technique have been addressed is the reliance of biomonitoring on epidemiology. As we discussed in chapter 4, epidemiologists use statistical methods to establish relationships between exposures and illnesses in and across populations. Environmental scientist Ken Sexton and his colleagues argue that "for epidemiologists to correlate environmental pollutants with health problems, they need to know who has been exposed and at what level. . . . Epidemiologists often find it difficult to establish cause-and-effect relationships for environmentally induced sicknesses. . . . Fortunately, there is hope: a method of accurately measuring not only contact with, but also absorption of toxic chemicals from, the environment—human biomonitoring." They go on to suggest that biomonitoring "provides unequivocal evidence that both exposure and uptake have taken place" (Sexton et al. 2003:38). Their article paves the way for concerted scientific partnerships; they also assert a place for the tools of public health in biomonitoring and a place for the technique of biomonitoring in regimes of public health and clinical practices. In their optimistic view, "a full screen of exposure biomarkers may be a part of every routine physical exam in the not-too-distant future" (45).

Perhaps the most striking recent example of the ascendancy of biomonitoring as the "it" technology was its use after 9/11. In *Science*, Stokstad reported, "After the World Trade Center towers collapsed on 11 September 2001, the world was gripped by the search for survivors. Researchers at the Centers for Disease Control and Prevention (CDC) raced to address an additional concern: the exposure of rescuers to potentially toxic smoke from the rubble" (Stokstad, 2004:1892). Examining blood from 370 firefighters, researchers found some exposure but no "dangerous levels," thus negating the need for further tests at that point (and presumably freeing up rescuers for other tasks). Environmental health scientist Paul Lioy and his colleagues claim that human exposure science, of which biomonitoring is a critical part, can help researchers better understand the environmental and health consequences of the terrorist attacks on the World Trade Center. They define human exposure science as "the study of human contact with chemical, physical, or biological agents occurring in their environments," and they assert that it "advances knowledge of the mechanisms and dynamics of events either causing or preventing adverse human health outcomes" (Lioy et al. 2006:6876). Not only was biomonitoring useful at the "complicated crime scene, war zone, and rescue operation" of Ground Zero, but it may also be useful for formulating "on the ground" strategies of containment and investigation of future disaster-related incidents and their long-term chemical consequences. Biomonitoring is thus central to public health infrastructures, as in the California case described earlier in this chapter, as well as to the emerging post-9/11 security apparatus.

In sum, biomonitoring represents ever more specific mechanisms to measure the body and assess relative health and risk of the body's behaviors and exposures. As scientific methods and technological innovations become increasingly adept at dissecting the body into knowable pieces of

"vital" information, we become more dependent on these corporeal frag-
ments as harbingers of our health on the one hand and of our doom on
the other. It is clear that our fluids, as taken from us and used for evalu-
ative purposes, are unpredictable and leaky. They become useful to others—organizations, the state, and forms of governmentality—and then are
made to tell certain kinds of biopolitical truths about our bodies. Fluids
and tissues come to matter more than the entire organism. These parts,
then, as if in some science fiction dystopia, achieve a monstrous power
over us in the form of numeric codes. Our bodies are no longer seen to
matter as individual lives, but, rather, our fluids and tissues taken together
represent a kind of environmental aggregate that speaks volumes about
the future of our species on the planet. Environmental activist Kristin
Schafer claims, "Information about chemicals in our bodies brings otherwise remote policy debates to a very personal level" (2004). And therein
lies the problem. For as journalist Harvey Black writes in *Environmental
Health Perspectives,* "How does one craft messages about the presence of
environmental chemicals in people's bodies?" (2006:A654). That is, once
we have this information about bioaccumulative risks, what can we do
about it, and what does it tell us about the gendered bodies from which it
came?

Liquid Gold: Or, Is Breast Milk Best?

If human beings are the top link of the food chain, then breastfeeding
infants are at its very apex. Yet this fact remains little recognized or discussed, except by an odd, small assemblage of mothers and scientists.
Ecologist Sandra Steingraber, a mother of two, laments, "I have yet to find
a poster or textbook that places a picture of a suckling child at the pinnacle of a human food chain, one full link above adult men and women. The
reason for this omission eludes me. . . . In any case, a failure to acknowledge the unique position of the breastfed infant within the ecological
world prevents us from having an informed public conversation about a
very real problem: the biomagnified presence of persistent toxic chemicals
in breast milk" (2001:251). Writing in the *New York Times,* journalist and
new mother Florence Williams admits, "Not only was nature's purest food
tainted by chemicals, but the act of breast-feeding itself, an act of love and
nurture, was also now marred by fear. Had I been wrong to be so smug
about the superiority of breast-feeding?" (2005:21). Native American activist Winona LaDuke and members of the Indigenous Women's Network

have made breast milk contamination a cornerstone of their environmental activism.[9] And a group of mothers in Northern California was so upset by news that rocket fuel and flame-retardants had found a way into women's breast milk, they started a grassroots organization called Making Our Milk Safe (MOMS). Founder Mary Brune commented in *Grist*, "There should be nothing more basic than a mother's right to provide clean and healthy breast milk for her child" (Dicum 2006). While these positions may be read as a calculated move because of the exalted place of breast milk in our culture alongside the emotional pull of cute, "innocent" babies, they also reveal something about the locus of toxic exposure and women's maternal concerns.

Evidence from biomonitoring has been used to show that "chemicals are still seeping into mammaries across the country" (Dicum 2006). This has both paved the way for new social collectivities such as MOMS and solidified the place of breast milk as an especially ideal fluid for measuring toxins in the body.[10] Breast milk has become symbolically and materially a potent site of pollution in the body, amplified by the breast's starring role in infant feeding. Toxins accumulate in our bodies by attaching themselves to fatty tissues, "which can be the same fatty tissues that hold the nutrients found in mothers' milk."[11] Women are considered to be at greater risk from toxins overall given the higher fat content of women's bodies vis-à-vis those of men (Vesely 2003), with breasts being one of the principal sites of accumulation. Moreover, breast milk is one of the few "exit routes" for chemicals and may contain higher concentrations of persistent toxic chemicals than other fluids or tissues (Steingraber 2001). The presence of chemicals in breast milk is troubling because of the consequences for women, in terms of risk for breast cancer and other diseases, but also for infants who receive the off-loaded toxins with their mothers' milk.

A number of scholars have drawn attention to the cultural politics of the breast (Yalom 1998, Jacobson 2000) and breastfeeding (Blum 2000), and we will not repeat their excellent arguments here. This chapter is not principally about the debate regarding infant feeding, although we tend to support the position that women should be enabled to breastfeed if they choose to do so but they should not be coerced into any one form of feeding their babies. Rather, here we are interested in breast milk biomonitoring for what it can tell us about uses of the female body, and for the ways in which it animates in especially visible ways social cleavages such as those centered on gender. Debates about biomonitoring of breast milk are messy, we suggest, because our cultural perceptions of breasts, breast milk, and infant feeding are continually in flux. As women's status in Western society changes, so, too, do collective notions of women's bodies and the "proper" way to feed modern babies, increasingly conceptualized as consumers. We traffic here in multiple cultural meanings and tensions underlying the catchphrase "Breast Is Still Best," showing how breasts are positioned as ideal resources for both infants *and* practitioners of biomonitoring. As a "tool for public health research," in the words of participants at a 2002 workshop on human milk surveillance, breast milk is positioned to circulate in and through numerous social and scientific worlds.[12]

So with respect to biomonitoring, what properties make breast milk "the right tool for the job" (Clarke and Fujimira 1992)? Notably, breast milk has a higher lipid content than blood serum; lipids are fat-soluble molecules. This makes human breast milk a valuable and nutritious source of food for developing infants. Yet because many persistent industrial chemicals are lipophilic—that is, they are attracted to fats—these categories of chemicals tend to bioaccumulate in breast tissue and milk. Some of these chemicals include pesticides, fungicides, and insecticides; PCBs (polychlorinated biphenyls) and PPDEs (protoporphyrin dimethyl esters); volatile organic compounds (VOCs); nitrate, nitrosamines, and nicotine; metals and trace elements; caffeine and drugs. The presence of chemicals in breasts and breast milk suggest that while the discourse of biomonitoring may emphasize total body burden, it is specific body parts that may be the most polluted. Widespread concerns about toxic breasts may lead to, among other things, reduced breastfeeding rates. After all, what new mother wants to dump a load of toxic waste into her soft, fresh, sparkling clean newborn?

This issue was framed by participants at workshops on human milk surveillance, as the potentially detrimental effects of biomonitoring on women's willingness to nurse their infants. Environmental risk analyst Judy LaKind and her physician colleagues state, "The panel emphasized the importance of unique ethical issues in human milk research, including the need to make special efforts to preserve and maintain breastfeeding rates" (2005a:1686). They report that only 17% of American women breastfeed, despite the widespread (some might say coercive) cultural and medical message that "breast is best"—that is, better than infant formula. Fewer than 5% of women in the United States breastfeed for a full year, which is the American Academy of Pediatrics recommended duration. The World Health Organization recommends that women nurse their babies for at least two years. While controversy continues to swirl around the heated issue of breast milk versus formula—often with tired new moms caught in the middle—most scientists, physicians, and women's health advocates assert that even polluted, human breast milk is healthier for infants than formula. Sandra Steingraber states, "The question is not whether we should feed our babies chemically contaminated, yet clearly superior, breast milk or chemically uncontaminated, yet clearly inferior, formula. The question is, what do we need to do to get chemical contaminants out of clearly superior breastmilk?" (2001:276).

The answer to that question increasingly hinges on biomonitoring, for which breast milk is considered ideal because large amounts of specimen can be obtained, evoking industrial metaphors of bovine milk production. Also, according to advocates, samples can be collected serially—that is, sequentially over time from the same women. It is worth noting that women only lactate after pregnancy and birth, and there is in fact typically only a limited amount and some women even "dry up." Yet environmental health scientists Paustenbach and Galbraith claim, "Samples of breast milk are relatively easy to collect" (2006:1146). We are disturbed by the notion that such material is "easy" to collect or that the techniques are noninvasive, given that breast milk must be harvested from the bodies of lactating women (and taken away from nursing infants).[13] In the material we reviewed, including scientific studies and media coverage, we found little mention of potential difficulties in securing milk specimens from women. Indeed, breast milk is discussed in this literature as a discrete substance with a life of its own, and women's whole bodies—the source—are rarely represented. The term "breast milk" itself is routinely abstracted from actual women. For example, toxicologist Suzanne Fenton and her colleagues write, "Human milk allows for a relatively long window of opportunity for specimen collection. . . . Large volumes of milk can be obtained using noninvasive collection procedures (although collection of milk from a new mother may be considered burdensome)" (2005:1707). Note the parenthetical nature of the final comment, as if this is not a truly significant issue.

We found just one article that discussed methodological issues in collection of human milk, based on material presented at the 2004 symposium on human milk surveillance. Pediatrician Cheston Berlin and his colleagues write in *Journal of Toxicology and Environmental Health*, "Ease of sampling and requiring the smallest possible volume of milk are the two most important factors in facilitating the collection of human milk for biomonitoring studies" (2005b:1811). Again evoking mass milk production, they go on to describe several key issues: sample volume, sample collection, potential diminution of rates of breastfeeding, storage needs to retain integrity of samples, strategies for disseminating findings of breast milk surveillance research, and chemicals for which analytical methods exist. The authors conclude by stating, "Investigations of women's attitudes to, and experience of, participating in milk collection studies could lead to information/recruitment methods/research practices that would encourage greater participation rates and make participation easier if that research were widely disseminated through breastfeeding advocacy organizations . . . and in the popular press" (1824). In this quest for greater participation among women, where are infant bodies and their nutritional needs located? After all, taking breast milk from women deprives infants. The "discovery" of breast milk's useful properties for risk assessment appears to lead to erasure and displacement of nursing infants, who are conceptualized simultaneously both as competition for the research material and as innocent victims of potential toxic exposure. Certainly one strategy is that researchers could recruit women who were breastfeeding and then switched to the bottle for the baby, yet might agree to continue lactating to be in biomonitoring studies.

From a biopolitical perspective in which regimes of surveillance are paramount, perhaps even more critical than collection issues may be the role of human breast milk as a sentinel fluid. For Commonweal's Sharyle Patton (as for many other environmental health activists), breast milk is "an indicator of what the community is exposed to and how exposed we all are."[14] While some commentators caution against generalizing from human milk data given that only some lactating women are represented in samples, others see biomonitoring as critical to "understanding the long-term health implications for the breastfeeding mother *and perhaps the general population*" (Wang et al. 2005; emphasis added). The rationale for breast milk biomonitoring in other countries, such as New Zealand and Germany, is precisely the ability to not just track women's and infant's exposure but to assess risk in and across communities spatially located. In human breast milk surveillance, then, this "most precious natural resource" serves as a sentinel for evaluating risk in the entire species. This is a potent example of the use of women's reproductive bodies for the collective benefit of humanity. Echoing some critical environmental studies of feminists (e.g., Stein 2004), we suggest the nursing woman and her presumed "natural" connection to the Earth renders her, problematically, a "mother" to the whole world.

Biomonitoring and Breast Cancer: Or, What's in a Mammary?

Might biomonitoring actually be beneficial to women or to babies and men as claimed? It could be useful to know what toxins are lurking in our bodies, especially if the partnership between biomonitoring and epidemiology can provide better data about actual health risks over the life course. Regarding one disease in particular, breast cancer, human milk biomonitoring has become especially salient. For example, the Breast Cancer Fund, a San Francisco-based nonprofit organization, includes biomonitoring legislation among its Internet resources and is an active participant in efforts to identify and eliminate environmental risks of breast cancer. And in popular media, the association of diseased breasts and chemical contamination has been made repeatedly. Indeed, there are distinct pro-breast milk groups (e.g., La Leche League) and environmentalists (e.g., Environmental Defense Fund) who use biomonitoring as an "issue" to lobby for their own concerns and agendas.

Logically, it is not such a leap to assume that the presence of poisons in one's breasts could potentially lead to cancer of said organs. Yet data are mixed. Environmental health scientist Richard Wang and his associates state, "it is not clear that measurements of chemicals in human milk are any more relevant to breast diseases or disorders than to diseases involving other organ systems in the body" (2005:1773). In contradiction, toxicologist Suzanne Fenton and her colleagues state, "Human milk biomonitoring may provide information on exposure trends in selected areas of the United States and may help to identify the agents most likely to alter breast development or sensitivity to other environmental carcinogens that affect the breast" (2005:1692).

Let us return to the California biomonitoring controversy to see how biopolitics of the breast played out in unique, sometimes surprising ways. An initial sponsor of the 2003 legislation introduced by Senator Ortiz was the San Francisco–based advocacy organization Breast Cancer Action. The organization, deeply interested in any links between chemicals in the environment and breast cancer, later withdrew its support out of concerns that a statewide biomonitoring program would discourage breastfeeding and thus negatively impact breast cancer prevention. There is, in fact, substantial data that breastfeeding may be protective against breast cancer.[15] In withdrawing its backing, Breast Cancer Action specifically objected to the idea that breast milk samples offer any valuable information about breast cancer risk: "Despite what may be the intuitive notion that the toxins found in breast milk are related to breast cancer, the pathways for the development of breast cancer may be (and probably are) very different from the pathways by which women accumulate toxins in breast milk."[16]

Breastfeeding advocates, such as La Leche League, shared some of Breast Cancer Action's concerns. A primary worry expressed by La Leche League was that media coverage of biomonitoring and body burden studies would scare women away from nursing their infants. Changes were made to subsequent bills mandating breastfeeding education as part of any biomonitoring program—and were a significant part of the bill signed in 2006—but tensions continued. Historian and author Christine Gross-Loh wrote in *Mothering* magazine, "While breastfeeding advocates fear that media misrepresentation leads many mothers to stop breastfeeding, environmentalists have wondered if breastfeeding advocates are not concerned enough about the very real problem of toxic body burden" (2004). Most breastfeeding advocates continued to recommend breastfeeding as the best food source for infants while acknowledging that toxins are present in breast milk and that more research needs to be done. The slogan "Breast Is Best" morphed into "Breast Is Still Best," with the added assertion that "environmental pollution is a reason to get rid of toxins from the environment, not to get rid of breastfeeding." Because the breast is not the exclusive location of toxins, these can be measured in alternative ways, lending credence to the claims of breastfeeding advocates.

The successful 2006 biomonitoring legislation was substantially revamped from the original 2003 language and subsequent iterations, in part because of concern by breastfeeding advocates and breast cancer organizations. The most striking change, a direct result of these concerns expressed through legislative testimony, was provision for an education and outreach program to encourage breastfeeding. Part of the outreach effort will be convincing women that biomonitoring their breast milk may ultimately make their milk and that of other women healthier by identifying toxins and working to eliminate them. Thus, we argue that the California biomonitoring legislation reproduces the widespread cultural claim that "breast is best" for babies without at all investigating or articulating whether breastfeeding is best *for women*. As with human milk surveillance writ large, women as people are absent; lactating mothers are represented most strikingly as tools and resources for securing healthy babies and a healthier population.

A concern related to gender, race, and class identified by only a handful of the participants in the California biomonitoring debate is the issue of coercion. If breast milk is understood to be representative of the toxins present not only in women's bodies but also in entire communities, the danger exists that, at some point in the future, biomonitoring may become mandatory rather than voluntary. Or, if not mandatory, then perhaps commoditized such that low-income mothers can earn money by selling their breast milk to institutions studying environmental health risk or to "pure milk" banks. The irony, of course, is that women of color and low-income women are already differentially impacted by breast cancer, including higher rates of mortality. We ask: What role does breast milk biomonitoring have in a society organized hierarchically along lines of gender, race, and class? Are there assurances in place that the communities most affected by pollution—largely low-income communities of color—will be those best served by the results of biomonitoring? Or will the concerns of the poor to live in clean neighborhoods be eclipsed by the desire of white middle-class breast cancer advocates living in Northern California and elsewhere to prevent and cure a frightening disease? In other words, is breast cancer the real point of research using biomonitoring, or are there other, more invisible agendas and concerns at work?

In short, diverse and even competing meanings are attached to breast and breast milk in the realm of human biomonitoring. For some breast cancer advocates, breast milk is toxic and thus a risk factor in developing cancer; for others, the act of breastfeeding may prevent breast cancer. For breastfeeding advocates, breast milk is "nature's choice," the superior food for infants and also culturally an important act of women's embodied agency and maternalism. Almost everywhere in the United States, from the American Academy of Pediatrics to La Leche League to popular magazines, we see claims that "breast is best." Some advocates suggest that breastfeeding is a basic human right: that mothers have the right to produce and infants have the right to drink pristine, pure, unpolluted breast milk. In this framing, biomonitoring is positioned uncritically as an essential tool to ensure that women's bodies are clean and pure—an ideal state that may be a long time coming. While we support the position that breastfeeding is a human rights issue, we also have to ask: Where are the men in the frameworks we have been discussing? Why is there so much emphasis on breasts and their milky discharge as essential for reproduction of the human species and so little attention paid to another generative fluid, semen?

The Duct Less Traveled

Environmental health scientists Paustenbach and Galbraith recently wrote, "Semen has not yet been widely used, but it is believed by some experts that semen may turn out to be a better test medium than many other bodily fluids currently being evaluated" (2006:1146). Indeed, a rudimentary form of biomonitoring of semen has been in existence for at least 50 years through the sperm banking industry, which connects testing of sperm to commodification. That is, semen has been tested as individual samples in the context of medical and commercial sperm banking. As a consumer product, semen is analyzed for exposure to a wide range of infectious and noninfectious diseases: HIV, hepatitis B, hepatitis C, human transmissible spongiform encephalopathy including Creutzfeldt-Jakob or "mad cow" disease, 7 treponema pallidum (which causes syphilis), human T-lymphotropic virus (types I and II), chlamydia trachomatis, and neisseria gonorrhoeae.[17] But this type of biomonitoring is used to measure the physiological health of singular sperm donors on behalf of prospective buyers, not to assess the ecological health of the planet. (To our knowledge, eggs are not biomonitored at egg banks, nor is breast milk routinely biomonitored at breast milk banks.[18])

While no sperm banks routinely check for *environmental* exposures or toxins harboring in sperm, banks do see the potential for "fertility insurance" for men to bank semen before potential exposure. Consider the following statement on the Internet site of the California Cryobank, one of the largest and most successful sperm banks in the country: "Our current environmental crisis has also generated a need for sperm bank services. Men who work in industries where there is the danger of exposure to radiation, toxins or other genetically threatening environmental pollutants are using sperm banks to preserve their sperm as insurance against possible accidents that could leave them infertile, impotent, or genetically damaged."[19] Increasingly over the past decade, epidemiologists and reproductive endocrinologists have studied the effects of exposure to a variety of environmental toxins on sperm count, sperm morphology (shape of sperm cells), and motility (movement of sperm cells) (e.g., Selevan et al. 2000, Ong et al. 2002, Sokol et al. 2006). Overwhelmingly, these studies are about the relative health of the sperm and the damage to its shape or speed rather than, as contrasted to breast milk biomonitoring, the use of the data as sentinel information for the entire human species. Research on biomonitoring and sperm appear largely to be about individual male reproductive health—"the actual health of the sperm for sperm's sake"— rather than about developing aggregate measures for assessing collective exposures.

Instead, semen and sperm have become targets and tools of biosurveillance used to trap dangerous men (e.g., sexual predators) but not research material for biomonitoring to examine men's overall toxic load. Semen is not collected in the spirit of communal good or to benefit mankind writ large; it is only located and collected after a crime in the spirit of social protection from individual men perceived as threatening or criminal. Unlike women whose bodies are often thought of primarily as vessels for reproduction, men are not typically seen as a collective biological resource—a fountain of seminal fluid that can be tapped for the broader social good. Rather, men's issue, their semen, is somehow represented as precious in the strangely discordant circumstances of being sold on the market to produce babies and, if spilled, to be used in ways that prove something nefarious has happened.

As explored in Lisa's book, *Sperm Counts,* semen is imbued with extremely intense social meanings including birth, death, disease, virility, sex, violence, love, hatred, and genetic heritage. Semen, of course, comes from the male body; it is constantly produced throughout the lifespan of a man, and it is ejaculated across most of the life course, intentionally and accidentally. It is conceptualized as a strong and tenacious fluid, always driving toward its goals. Further, semen as a material object can endure in vaginas and anuses and on fabrics and surfaces. While semen is collectible, it is not easy for the ejaculator or his partner(s) to catch unless there are prior arrangements made to entrap it, such as available mouths, vaginas, anuses, hands, beakers, or condoms to take it in. Therefore, it can be messy and unpredictable, often leaving recalcitrant traces for others to find.

Not only do people who are recipients of semen have a difficult time getting rid of it, when spilled it leaves an evidentiary trail that can lead— with the proper technologies and criminal-scientific investigatory apparatus—to just one, specific man. By establishing methods of seminal detection, professional and lay forensic practitioners are able to locate semen, describe the type of sex men and women (or men and men) are performing, and ultimately substantiate an act of intercourse. For example, in 1953 biochemist Paul Leland Kirk published *Crime Investigation,* one of the first texts on criminalistics and crime investigation that encompasses practice as well as theory. Building on previous methods for collecting evidence at crime scenes, Kirk likened biological fluid to a "silent witness," as he called it. This silent witness, the biological "leak," reveals specific information about the individual and his actions: "Wherever he steps, whatever he touches, whatever he leaves, even unconsciously, will serve as silent evidence against him. . . . The blood or semen that he deposits or collects—all these and more bear mute witness against him" (Kirk 1953:4).

Sperm is thus a sneaky and leaky silent witness. Once (and still) both coveted and assumed dangerous because of its ability to impregnate via penetration of an egg, through biotechnological innovation semen has also become the equivalent of a mobile identifier. A man can be identified and defined, and perhaps held accountable, by the actions and trajectories of his sperm. With both the means and the authority, the state now can codify sperm cells, making sperm and the individual men who produce it inherently knowable and definite. The state can offer us mechanisms to make us feel safer through the codification and translation of data, but there are trade-offs: the loss of privacy and criminal profiling using a public database of "dangerous" men that is subject to bias (racial and otherwise) in its construction and to abuse in its deployment.[20] Seminal fluid becomes a proxy for actual men's bodies, and in some ways is even more valuable than the originating bodies in that fluids can prove certain things, such as rape or intercourse, happened. Who or what the actual body is becomes almost irrelevant once the evidence is extracted, processed, codified, and marshaled in a courtroom.

We want to draw attention here to the very different ways in which sperm is monitored and deployed vis-à-vis breast milk. While sperm is assessed for what it can tell us about an individual man and his health and actions, and whether these need to be addressed through treatment or prosecution and incarceration, sperm is not yet widely used to measure or make predictions about the eventual demise of the species.[21] There has been some alarm regarding threats to "male fertility" that could be read in terms of species risk (Daniels 2006), yet overall sperm is considered "sentinel" only to the man from whose body it issued (assumptions *are* made about the racial categories into which individual men are placed). Breast milk, in contrast, is made to tell us less about women's individual health;

data from women's bodies is used to calculate infant and species risk. Women are the invisible bodies yet highly potent tools through which toxins flow and data are transmitted, and they are made visible through their corporeal fragmentation. Fluids and tissues may be extracted from women's culturally "fertile" bodies (regardless of whether individual women are fertile or infertile, they are presumed to be always potentially *reproductive,* as discussed in chapter 3), and made to speak about broader collectives such as families, communities, nations, and even the planet. In highly traditionally gendered ways, men and their parts are conceptualized as threatening and violent, inherently dangerous, while women are framed as mommies to and of the Earth who must selflessly sacrifice their precious fluids for the health of the collective. It is not so much, we argue, that biomonitoring technologies do not yet exist to process sperm but that our cultural meanings about gender and the bodies of women and men have not evolved in ways sufficient to facilitate use of sperm as a sentinel fluid. Fluids may matter, to be sure, but it appears that gender—and especially women's inescapable connection to fetuses and infants—may matter more in determining whose bodies are enrolled to speak for the species collective.

Reconceptualizing "Body Burden"

As we argued in chapter 4, HIV-positive pregnant women are positioned as sentinels in epidemiological research. Continuing this theme, we have demonstrated in this chapter still another way in which women's reproductive bodies and functions are manipulated and made invisible—often with women's maternal complicity—in service to the broader collective, beginning with infants at the pinnacle of the food chain. Drawing on the term "body burden," in wide use in the social worlds of biomonitoring, we want to suggest a critical reconceptualization. As discussed earlier, body burden describes the chemical load in individual bodies, which is then aggregated in the case of lactating women especially to assess species load and overall risk. We assert a different use of body burden here, both theoretically and politically: Whose bodies bear the burden, not just of bioaccumulative toxins but of the weight of the biomedical-industrial complex, scientific technologies, and environmental risk assessment strategies? Whose bodies are reduced to data, in what ways, and with what consequences? We suggest that while all human beings may be exposed to hazardous waste in our earthly habitats, only certain human bodies are exposed to new knowledge-making institutions and practices, such as biomonitoring. Surveillance may be proliferating and expanding, but it is doing so in deeply stratified ways.

Biomonitoring is kin to epidemiology and demography; all represent forms of embodied knowledge central to the formation of new, rationalized subjects in the 21st century. As such, there is a lack of critical engagement with ethics; advocates and users of biomonitoring may bulldoze through this "objective science" that provides us with meaningful data, just as they mine fluids and tissues for manipulation in a laboratory and frame collection issues as merely technical. What we want to insist on—to reveal analytically—is what gets lost in the aggressive scale-up of biomonitoring and how particular bodies are used and deployed for other purposes. For example, what benefit exists in the knowledge that our bodies are polluted, if we do not have the resources or capacity to change the social conditions that facilitated the exposures in the first place? While tools of biomonitoring are useful in environmental justice struggles, as sociologist Rebecca Altman (2008) shows, how might such data be used in more insidious ways?

Clearly, while we, along with Winona LaDuke and her comrades (Associated Press 2003), would like to know if Native American women are breastfeeding their children toxic breast milk because they were once relocated to a Superfund site, we also worry that biomonitoring data is not necessarily used to help these women. Rather, such data may be used by corporations to muddy the waters of paths of exposure, to protect themselves regarding historical dumping of toxins on indigenous lands, or to sell baby formula. Perhaps just as the HIV/AIDS-infected bodies of dehumanized Africans are erased but used to provide valuable information for people in the West, toxins are contained and measured in human guinea pigs spatially distant from "us," such as women on Native American reservations. In the realm of data collection from human body parts, whose bodies are subject to such biosurveillance and who benefits? Is the developing fund of knowledge based on biomonitoring techniques used, in fact, to help the collective, or is it used to measure the collective as a way to develop strategies that may reproduce the very hierarchies responsible for exposure?

In sum, the process of extracting material from human bodies, while framed as useful for the social body writ large, is not done democratically. Even when new mothers may consent to participate in breast milk biomonitoring, these techniques rely on extant stratification systems and pervasive cultural ideals about gender, especially maternal sacrifice. A core idea of biosurveillance is that knowledge may enable risks to be minimized, as discussed in chapter 4 and here. But risks for and to whom, and from what? Regarding semen, risk is "minimized" only after the fact of semen's collection and use; that is, there is no preventive measure taken against sexual assault or pandemic male violence. In this case, fluid collection facilitates proof and punishment, a very limited form of "social good," rather than broader protection of children and women. Breast milk may be used to predict species risk, but biomonitoring of women's bodies is only peripherally framed or practiced as a women's health issue. We suggest, then, that whole bodies are relegated to the margins here. We use fluids and tissues in a kind of politics of extraction and abstraction that reassure us something is being done about "the environment" and "crime," while also reaffirming our beliefs about bad men and good mothers.

As we move into the next section of this book, focused on heroes, we find a different set of practices that are also deeply gendered and rely on the gendering of women's and men's bodies in the competitive, consequential worlds of war and sport.

PART III

Heroes

In this last year of American history, we witnessed acts of sacrifice
and heroism, compassion and courage, unity and fierce determination. . . . Children reflect the values they see in their parents and
in their heroes, and this is how a culture can be strengthened and
changed for the better.

> George W. Bush, *Remarks of the President on Teaching
> American History and Civic Education Initiative* (2002)

American identity and fervent patriotism are inextricably
linked with masculine heroism. Our national consciousness is rife with
notions of victory, triumph, valor, bravery, and winning, usually at all
costs. Gary Cooper, John Wayne, Clint Eastwood, Tom Cruise: these are
our cinematic stars, just as Ernest Hemingway, Cormac McCarthy, and
Larry McMurtry are our quintessential storytellers. The rugged, solitary
man who is larger than life, questing for glory after overcoming tremendous adversity, is the archetypical American. And better yet, if this hero
can be trademarked, our late-capitalist machine can accessorize and mass-
produce this heroism for willing consumers and spectators. Sell more
tickets, or in the case of macho stars like Arnold Schwarzenegger, become
governor. Just like 1980s pop songstress Bonnie Tyler, we're all "holding
out for a hero," and we're willing to pay for it.

Stories about heroes work in many ways because we are taught from an
early age to follow the narrative arc. From Greek literature to Harry Potter, all the clues are in place for the savvy consumer to determine which
character to root for. Heroism tales provide cues and mechanisms that
enable an audience to identify with the hero and usually introduce villains or arch enemies in a wholly good and wholly bad dichotomy. Tales
of heroism are not simply enjoyable; they comprise the fabric of our

social conditioning. Little wonder that as the status of girls and women has changed, we have clamored for more female heroines.

Through the American innovation of the superhero, children and adults alike are taught to empathize with characters that embody superhuman qualities. Bruno Bettelheim's *The Uses of Enchantment* (1989 [1975]) explores the creation, dissemination, and consumption of fairy tales as processes of socialization. To read fairy tales to children is to transmit values in preparation for social rituals, such as being a good citizen. But children's literature does more than convey cultural norms. Sociological and historical analyses of many texts have revealed persistent and pejorative representations of subordination, inferiority, weakness, and stupidity for those who rank low on the social ladder. Gender, racial, ethnic, age, and ability biases have been embedded in children's literature since the beginning of the genre. Sex and gender role stereotyping is perhaps the most prevalent form of discrimination in preschool children's books.

Turning to the theater of war, a potent site of myth making, we are interested here in two countervailing war stories: Jessica Lynch and Lance Armstrong. We use the term "theater of war" deliberately to denote the spectacular aspects of how each of these warrior biographies has been narrated. War produces missing bodies, both literally and figuratively, while amplifying the visibility of other bodies. Since the advent of photography, wars have been characterized by iconic images of heroism and victory alongside bodies rent horribly by violence. For example, pictures of mass graves from the Holocaust still haunt us. The infamous image of Phan Thị Kim Phúc, the nine-year-old Vietnamese girl running naked down the streets burning from napalm, has been seared into our consciousness. And more recently, many of us have been horrified and shamed by images of U.S. soldiers torturing prisoners at Abu Ghraib.

The Iraq War itself may be considered spectacle. Its motives and origins have been under suspicion since its inception. The public was offered shoddy evidence to encourage their support of the war at a time of mass vulnerability and fear due to the terrorist attacks of 9/11. As spectacle, just as the Gulf War (1990–1991) emphasized disembodied smart bombs and other cutting-edge weapons, the current conflict also has erased the dead. It is very difficult to get information about the dead, missing, or injured—of the military or especially of civilians. Despite many reporters being embedded with U.S. troops, the media has been neutralized. Mainstream media, in particular, has been complicit with the Bush administration in offering a sanitized narrative of this war and the conflict in Afghanistan.

What has appeared alongside critical analysis of this misguided military engagement are fairy tales complete with ogres, a princess to be rescued, valiant warriors, fallen heroes, and very clear demarcations between good and evil. The popular culture tales of war shadow critical accounts. Any deviation from the script, such as the systematic harassment and rape of women soldiers by their male comrades, is written off as individual pathology and not the outcome of armed conflict and institutionalized and militarized violence. (For an excellent feminist critique, see Oliver 2007.)

Two recent mythic characters in the staged war chronicles are rescued prisoner of war (POW) Jessica Lynch and seven-time Tour de France winner Lance Armstrong. Armstrong's story works because it has a willing hero, eager to be trademarked and exploited for his heroism: Lance Armstrong™ conquers cancer, the world, and the marketplace. Initially, Lynch's story also works, but in the end, her own narration interrupts the neatly constructed script. Truly gendered tales, Armstrong the white knight rescues a nation and gives us hope of victory, while delicate Lynch is rescued from the "axis of evil," like the princess in the tower, by patriotic American soldiers.

And this, in our view, is where the heroism arc gets really interesting. When Lynch returns to tell her story, when she is again fully conscious, she talks back to The Man: "I think this story has been manipulated to sell the war." Debate ensues in which people argue about whether she is indeed a hero. There is no disagreement that saving her was heroic; the heroism of the act of rescue is taken for granted. The soldiers who retrieve her are indisputably heroes. But because the princess speaks out against the King, loyal patriots demand that she return the Bronze Star. Everybody knows that the mute princess isn't supposed to have an opinion.

Lynch, who practices free speech, is seen as an enemy of freedom, the war, and the president. Bad hero. Bad woman. Bad princess. Bad girl. Bad American. Comparatively, Armstrong seems to be writing his own script as the Good American. He is the quintessential speaking subject, a multimillion-dollar, true blue, patriotic embodiment of male privilege and national pride.

Everybody may be holding out for a hero, but not every body can be a hero.

6

"They Used Me"
Manufacturing Heroes in Wartime

Consistent with the way in which the American news media represents casualties, the only bodies we can even imagine beneath this cloak of patriotism are American. If the only bodies that count as casualties of the war are those of American soldiers, often pictured as male, then one must ask, which bodies are obscured from our sight? Why are some bodies put on display while others are made invisible?

Grace M. Cho, *Gendered Bodies: Feminist Perspectives* (2007)

Strolling the aisles of any department store's toy section is a lesson in gender segregation. Miniature plastic and plush bodies of humans and animals—particularly babies—smile from behind their veneer of packaging. There is an aisle for boys where soldiers, construction workers, and racecar drivers populate the shelves and an aisle for girls full of Barbies, princesses, fashion models, and puppies and ponies. The boys' aisle is blue, green, red, brown, and orange; the girls' aisle is pink, purple, white, peach, and yellow. These toy bodies are highly visible, attractive, and typically garishly displayed to capture the child's gaze—stimulating a desire to play, to imagine, and to identify. The department store's strict codes are prophylactic against potential contamination by gender "cooties," that is, boys' "blue" toys stay on their shelves not touching the girls' "pink" things. Boys are encouraged to play at war, girls to play at homemaking and childcare. These aisles are sites of childhood socialization to the gender order; there is no legal barrier here to gender transgression (i.e., picking up the "wrong" toy does not result in arrest). The production of gender requires much more intensive ideological policing.

In longstanding rituals of socialization, little boys' bedrooms are often organized as elaborate staging fields of battle peopled with two-inch army-green toy soldiers. Such tableaus and remembered experiences have become part of many men's childhood nostalgia. The ever-popular G.I. Joe doll, recently having undergone some workouts at the gym, is still heavily marketed to boys. Today's G.I. Joe, originally introduced in less overtly macho form in 1964, is now a ripped fighting machine with "comblike muscles that run along the sides of his body" (Pope et al. 2000:42). Despite bulking up, though, G.I. Joe is more apt to be "played with" in the digital world or watched in the animated webisodes of Marvel Comics. Playing is now virtual, without the somatic connection to actual toys; rather, highly mediated interactive "screen-time" constitutes play. The bodies of the plastic action figures themselves, real toys such as G.I. Joe, are increasingly missing in action.

We are interested here in another live action figure in fatigues: Jessica Lynch. The slender, blonde, 19 year old from West Virginia enlisted in the army before 9/11, after conversations with a persuasive recruiter who explained the military's financial and educational incentives. The story of Lynch's 2003 capture and rescue led to two distinct media blitzes. The first in 2003 immediately followed the rescue mission in April and continued until the November Veteran's Day release of both journalist Rick Bragg's biography of Lynch, *I Am a Soldier, Too* and the NBC movie *Saving Jessica Lynch*. Front-page stories about Lynch ran in the *New York Times,* the *Washington Post,* and *USA Today,* and news programs on CNN and MSNBC featured her story for a week. She was on the cover of *Time* magazine and interviewed by Diane Sawyer as part of ABC's *Primetime* series, as well as a two-part interview on *The Today Show* with Katie Couric. She appeared on *Larry King Live* and *David Letterman.* The second media blitz, not nearly as glamorous, occurred in 2007 and is analyzed below.

But first let us establish what is "known" about Jessica Lynch. Lynch attended basic training at Fort Jackson, South Carolina, and Fort Lee, Virginia. After deploying to Kuwait, on March 23, 2003, Private First Class Lynch, a supply clerk in the 507th Maintenance Company, was injured in an "ambush" as her convoy drove from Kuwait on a supply run to Iraq. The company took a wrong turn near Nasiriyah and was attacked by Iraqi forces. Eleven of her fellow soldiers were killed, and five others were captured (Broder 2003, Faludi 2007). An injured Lynch was taken to Saddam Hussein General Hospital. Some sources claim it was Iraqi attorney Mohammed Odeh al-Rehaief who informed U.S. forces of

Lynch's capture and whereabouts. Allegedly, he walked six miles to a Marine checkpoint to alert them to Lynch's whereabouts, furnishing them with maps. He penned a book documenting his participation in Lynch's story, although Lynch and the physicians who worked at Saddam Hussein Hospital have disputed some of his claims (al-Rehaief 2003). For example, al-Rehaief and his supporters have alleged that his wife was a nurse at the hospital and that is how he had witnessed Lynch's location and treatment; however, there is no evidence of his wife's employment at the hospital. Other accounts state that it was al-Rehaief's sister-in-law who was a doctor at the hospital and that through her access, al-Rehaief was able to delay Lynch's leg from being amputated. There is also dispute about Lynch's treatment: she claims she was well treated; he claims she was slapped and this motivated him to get involved. Al-Rehaief's account is the basis of the 2003 NBC movie *Saving Jessica Lynch*. Although she did view part of the movie, Lynch was reportedly too disturbed by the factual errors in the movie to watch it in its entirety.

In 2003, al-Rehaief was granted asylum in the United States under humanitarian parole, a status typically awarded for urgent cases. He was as of this writing employed as a consultant with the Livingston Group, a conservative Washington D.C.–based lobbying firm. Despite the disputed story of al-Rehaief's involvement, there is consensus on other Iraqis' involvement in Lynch's rescue. Approximately one week after her capture, an Iraqi doctor, Harith al-Houssona, attempted to drive Lynch to a U.S. checkpoint to return her to the Americans, but GIs fired on the ambulance, which was forced to return to the hospital. On April 1, 2003, a team of commandos—a real-life integrated operating unit—comprised of U.S. Army Rangers, Navy SEALs, Green Berets, and Air Force pararescue jumpers stormed the hospital to rescue Lynch. They also recovered eight other bodies.[1] On April 2, 2003, within hours of recovering Lynch from the hospital, the Pentagon released a five-minute video of her daring "rescue" to reporters. Famously, the soldiers supposedly stated, "Jessica Lynch, we're United States soldiers and we're here to protect you and take you home," to which she is said to have replied, "I'm an American soldier, too" (Bragg 2004:131).

Airlifted to Germany's Landstuhl Regional Medical Center, the hospital's commander Col. David Rubenstein cataloged Lynch's injuries during a press conference.[2] These injuries included fractures to her right arm, both legs, right foot and ankle, and lumbar spine. She also sustained a head laceration. After several surgeries to stabilize her injuries, including spinal surgery, Lynch was transported to Walter Reed Army Hospital. On August 27, 2003, after three months in the hospital, Jessica was given a medical honorable discharge.

Almost immediately after her rescue, news reports, such as one from the *Washington Post* on April 3rd, claimed that Lynch "fought fiercely and shot several enemy soldiers . . . firing her weapon until she ran out of ammunition" and "she was fighting to the death, the official said. She did not want to be taken alive."[3] These media reports spread quickly, and Lynch's story became a multimedia spectacle from April through November. Lynch was awarded the Purple Heart for being wounded in combat, a Prisoner of War Medal, and a Bronze Star for meritorious service in combat. Yet, there were fissures in the officially narrated, military-ratified story circulating about Lynch. During the November 2003 interview with Diane Sawyer for *Primetime,* Jessica Lynch explained her belief that the U.S. military had overdramatized the story of her rescue. The media coverage of her rescue often bypassed Lynch herself, constructing stories that did not match Lynch's own experience. She stated, "It hurt in a way that people would make up stories that they had no truth about. Only I would have been able to know that, because the other four people on my vehicle aren't here to tell that story. So I would have been the only one able to say . . . I went down shooting. But I didn't."[4]

The second media blitz occurred four years later, in 2007, following testimony Lynch delivered to Congress. On April 24, 2007, during the Committee on Oversight and Government Reform hearings titled "Misleading Information from the Battlefield," Lynch stated, "it [her story] was under staged by media all repeating the story of the little girl Rambo from the hills of West Virginia who went down fighting. It was not true. I have repeatedly said when asked that, if the stories about me helped inspire our troops and rally a nation, then perhaps there was some good. However, I am still confused as to why they chose to lie and try to make me a legend when the real heroics of my fellow soldiers that day were legendary."[5] Later Lynch stated, "The truth is always more heroic than the hype." She claimed in multiple reports, and most significantly during her congressional testimony, that her gun jammed before she could fire, that she was well treated by hospital staff, and that she felt "used" by the U.S. military and media. Not nearly as popular as she was during the 2003 stories, in 2007 Lynch did appear on CBS's *The Early Show,* but there were no movies, book deals, or late-night talk show visits to showcase this particular act of bravery.

How might we understand this complicated action figure? A young woman who goes to war, is manipulated for larger geopolitical purposes, is misrepresented by the government, and yet stands up for herself before Congress, while simultaneously bringing the focus back to her fallen comrades? Dolls based on Jessica Lynch, and other women in the military, are not found on the shelves at Wal-Mart or Target—but their babies are. Possibly the lack of women soldier action figures is market-driven: in other words, there is no demand. But it may also be due to the fact that women soldiers are disabled from full participation in the military, either through official exclusionary policies regarding combat or everyday practices of intimidation.

For example, through practices of rape and sexual harassment exemplified by the 1991 Tailhook and the 1996 Aberdeen scandals, the military as an institution continues to exhibit hostility to fully integrating women into its ranks.[6] Aberdeen Proving Ground, the U.S. Army base in Maryland, was the site of sexual assault and rape of female trainees by their male superiors, 12 of whom were indicted.[7] The very materiality of women's bodies and their routine sexualization is clearly troublesome to the military (e.g., Williams and Staub 2005). The pain and suffering of women's bodies at the hands of the enemy is deployed symbolically as a means to bolster the military's masculinity and nationalism: "We can and will protect our women." But the actual bodies and experiences of women are missing in action when their pain and suffering is caused by the military itself.

Ironically, Iraqi bodies, and in particular Iraqi women's bodies, are ghostly in our own research on the war in Iraq and its consequences. Beyond the paradigm of rape and sexual violence, sexual access to women's bodies is a fundamental part of the institutional fabric of war. A 2003 Human Rights Watch report titled "Climate of Fear: Sexual Violence and Abduction of Women and Girls in Baghdad," established credible evidence of at least 25 cases of sexual violence and abduction in the city during the period between May 27 and June 20, 2003. The report states, "At one police station that Human Rights Watch visited, Iraqi police officers said that prior to the war they typically received one rape complaint every three months but had seen several cases in the few weeks it had been reopened since the war."[8]

Furthermore, the Iraq War has spawned other sexual practices that are not as coercive as rape but not entirely "consensual," given the violent structural conditions. Consider the experiences of many Iraqi refugees in

Syria and other countries. Reports from the United Nations and investigative journalists revealed the experience of Iraqi refugees engaging in "survival sex" (Clark-Flory 2007, Zoepf 2007). Here, women exchanged sex or erotic entertainment, such as stripping, for food and other basic necessities. Some U.N. estimates claim that 70–80% of prostitutes in Damascus are displaced Iraqi women. We suggest that Iraqi women (and men and children) became even more invisible by the distorted overexposure of Jessica Lynch's biography and other characteristically American stories.

"A Bitch, a Ho, or a Dyke": Women in/and the Military

During war, human bodies are thrown together in perhaps the most desperate of circumstances—bodies collide, limbs are severed, flesh is seared.[9] Domination and aggression are used to overpower and subordinate "enemy" populations; military colonization is an operation of masculinity, exploiting and pillaging natural resources including the bodies of women (Barstow 2001; Frederick 2000; Enloe 2000, 2007; Brownmiller 1993). The phrase "the spoils of war" is extremely gendered, whereby women's bodies are seen as an extension of territory to conquer and trophies to collect. Historically, overpowering women's bodies through rape is part of a larger, often sanctioned, violent practice of militarized colonization; pregnancies that result create a form of racial, ethnic, and national breach. As Canadian legal scholar Valerie Oosterveld writes in the *UN Courier* "shocking reports were published around the world about the use of rape and forced pregnancy as tools of 'ethnic cleansing' in Bosnia. . . . The same pattern is true of the former Yugoslavia, where women were raped until they were pregnant and then held until they were close to term. In 1993, it was estimated that between 1,000 and 2,000 women there became pregnant as a result of rape" (1998).[10]

Investigating the invisibility of reporting on sexual terrorism in Iraq, Ruth Rosen (2006), a historian and journalist, has written about the premeditated rapes and murder of 14-year-old Abeer Qassim Hamza and the killing of three of her family members by five U.S. servicemen from the 101st Airborne Division based at Fort Campbell, Kentucky, in March 2006. Three of the men have since pled guilty, another was convicted of rape and murder, and another was still awaiting trial as of this writing. Using this case as evidence of a larger epidemic of underreported sexual violence in Iraq, Rosen (2006) contends, "Like women everywhere, Iraqi women have always been vulnerable to rape. But since the American invasion of their country, the reported incidence of sexual terrorism has accelerated markedly—-and this despite the fact that few Iraqi women are willing to report rapes either to Iraqi officials or to occupation forces, fearing to bring dishonor upon their families."

We are concerned here with the ways that gender is deployed within the military and cast on the militarized body (Enloe 2000). In a 2003 interview with Salon.com, Spc. Mickiela Montoya of the National Guard stated, "There are only three kinds of female the men let you be in the military: a bitch, a ho or a dyke"; 23-year-old Jennifer Spranger was offered $250 by her team leader if she would give him a hand job. In her 2005 memoir, *Love My Rifle More Than You*, Kayla Williams, a translator and former U.S. Army sergeant, reports being offered money in exchange for showing her breasts to her male comrades.

Researching a book about American women soldiers who served in the Iraqi War, journalism professor Helen Benedict found a striking similarity in her interviews: "Every one of them said the danger of rape by other soldiers is so widely recognized in Iraq that their officers routinely told them not to go to the latrines or showers without another woman for protection."[11] During an interview with *Democracy Now* radio host Amy Goodman, Benedict discussed the case of three female soldiers who were terrorized by possible sexual assault and died of dehydration:

> AMY GOODMAN: And they were simply afraid to go out alone to get the water, being harassed?
>
> HELEN BENEDICT: Yeah. And she (Karpinski) told me—and this is something more detailed than came out in what you just showed—that there were men who were waiting out there, and they were pulling women into the latrines and abusing them and raping them there. And that's—word had spread about this, and that's why the women were afraid to go out. And I went to the site, the Iraq casualty site, which lists all the deaths, and I did indeed find three deaths of women in the year she was talking about attributed to non-hostile causes, which the Army never seems to really explain, so I think it's very possible those are three she was talking about.

To complicate the process of legitimizing as problems the abuses against women, whistle-blowers are often in fraught positions. Consider, for example, U.S. Army Colonel Janis Karpinski of the 800th Military Police Brigade and commander of Iraq-based military prisons in 2003 during the Abu Ghraib scandal. She was subsequently demoted to brigadier general for dereliction of duty yet has publicly stated that the abuse of prisoners was ordered by Defense Secretary Donald Rumsfeld and that she was demoted based on political retribution. In 2006, Karpinski testified at a mock trial known as the Bush Crimes Commission Hearings, a series of testimonials concerning the Bush administration's alleged crimes against humanity. The commission was created by the Not in My Name Project, a national network of individuals and organizations formed in 2002 to resist the U.S. response to terrorism.[12] Her testimony was based on the deaths in 2003 of three female soldiers; like Benedict, she contended that the women died from dehydration because they feared sexual assault by their fellow soldiers:

> COL. JANIS KARPINSKI: Because the women, in fear of getting up in the hours of darkness to go out to the portoilets or the latrines, were not drinking liquids after 3:00 or 4:00 in the afternoon. And in 120-degree heat or warmer, because there was no air conditioning at most of the facilities, they were dying from dehydration in their sleep. And rather than make everybody aware of that, because that's shocking—and as a leader, if that's not shocking to you, then you're not much of a leader—so what they told the surgeon to do was, "Don't brief those details anymore. And don't say specifically that they're women. You can provide that in a written report, but don't brief it in the open anymore."

The Army has labeled her charges unsubstantiated, but Karpinski, despite pressure from the military, still claims they are true.[13]

Troublesome female bodies, it would seem, while desirable, disappear in the military either through official attributions of their deaths to vague nonhostile causes (as in the case of the three women who died of dehydration) or through punishing these bodies and keeping them out of commission (as in the case of Suzanne Swift). On June 11, 2006, Army Specialist Swift was arrested from her mother's home and subsequently confined to her base, after being accused of going AWOL.[14] She had previously reported sexual assault and abuse by her commanding officers in Iraq and in the United States. In a September 2006 interview with Amy Goodman from *Democracy Now,* Swift reports on her situation:

> SPC. SUZANNE SWIFT: Well, the one that I tried to report was my platoon sergeant. And, you know, looking back now, I had a squad leader who literally singled me out to be the person that he was going to have sex with

during the deployment. And, you know, I did. I was nineteen. I fell for it, and for months I was like his little sex slave, I guess. It was disgusting and it was horrible, and I didn't know what to do.

AMY GOODMAN: And so, ultimately, what happened?

SWIFT: Ultimately, I stopped it. I told him that I didn't want to continue this relationship. And he made my life hell. I mean, a squad leader in the Army is basically—that's your boss. Everything that you do—eat, sleep, go to the bathroom, when you go to work, everything—they can tell you when to do it and how to do it. And he made my life miserable, because I wouldn't have sex with him anymore.[15]

The stories above may be shocking because they are stories that have not been officially told. They, like the bodies that narrate them, have been rendered officially invisible, delegitimated. The incongruent bodies of women in the military, particularly women who do not consent to sexual advances, are disciplined, as feminist scholars have shown. In books such as *Maneuvers: The International Politics of Militarizing Women's Lives* (2000) and *The Curious Feminist: Searching for Women in the New Age of Empire* (2004), political scientist Cynthia Enloe's distinctive contribution has been to unveil processes of militarization in which *gendered* national and global identities are formed and re-formed. Her books powerfully critique militarization as a practice that both creates and is fueled by the embodied enforcement of masculinity and femininity as nation-building activities. In Enloe's work, militarization is not just a *thing*, but it is, rather, a complex, stratified, consequential system of discourses, practices, and relationships in which gender and sexuality matter very much indeed.

Governments in Western societies maintain social order through masculine values and practices—coercion, domination, and military and police strength. Men's bodies swell the ranks of military institutions and the corporations, such as Blackwater and Halliburton, that benefit from the military—the "military industrial complex." The way war is narrated to those of us "back home" follows particular tropes and visual cues to capture the requirements of grit, physical endurance, and masculine bravado. Indeed, the "embedded" reporter, the war correspondent who reports from within the intimate theater of war, was a military public relations concept refined during the Iraq War (Katovsky and Carlson 2003). As journalist and novelist Zoe Heller (2003) writes in an essay critical of the American practice of embedding reporters, these reporters deliver the news "while riding shotgun on tanks, with the wind in their tousled locks." Heller expresses concern about the conflicts of interest that might emerge when embedded reporters rely on the military for their sustenance. She aptly points out the irony in the following quote: "'I can tell you that these soldiers have been amazing to us,' David Bloom, a reporter for NBC, traveling with the Third Infantry, reported the other night. 'They have done anything and everything that we could ask of them and we in turn are trying to return the favor.' Cute. But not the sort of thing you want to hear from a newsman."[16]

Anthony Swofford recalls in his 2003 memoir, *Jarhead: A Marine's Chronicle of the Gulf War,* his commanding officer provided specific instructions regarding how to display his body to the press during the 1991 Gulf War: "He's ordered us to act like top marines, patriots, shit-hot hard dicks, the best of the battalion. . . . 'Listen up,' Dunn says. 'I've gone over this already but the captain wants you to hear it again. Basically, don't get specific. Say you can shoot from far away. Say you are highly trained, that there are no better shooters in the world than marine snipers. Say you're excited to be here and you believe in the mission and that we'll annihilate the Iraqis. Take off your shirts and show your muscles. We're gonna run through some calisthenics for them. Doc John, give us a SEAL workout. Keep it simple, snipers.'" Embedded reporters appear to be seduced by staged performances of masculinity, and troops are made complicit.

In the tradition of the war correspondent, President George W. Bush, as part of the Iraqi War media carnival orchestrated by his administration, embedded himself within the military by dramatically landing in a Navy S-3B Viking on the deck of the USS *Lincoln,* where he infamously proclaimed, "Mission Accomplished!" Richard Goldstein's (2003) *Village Voice* article commenting on the May event draws our attention to Bush's attire, a fighter-pilot jumpsuit: "Discretion prevented anyone from mentioning that Bush's outfit gave him a very vivid basket. This was the first time a president literally showed his balls. . . . Clearly Bush's handlers want to leave the impression that he's not just courageous and competent but hung."

Journalist Naomi Klein (2007) critiques neoliberal practices of what she calls "national shock therapy" as part of collectively experienced terror where we are "inclined to follow leaders who claim to protect us." Even in peacetime, nations often behave as if they and we are under attack, which stirs up fear and anxiety and shocks their citizens, in turn justifying spending a large part of the national budget on criminal justice

President George W. Bush on the *USS Abraham Lincoln* at sea, in May 2003, after landing in a U.S. Navy S-3B Viking jet piloted by Commander John Lussier, shown on right. The president declared "Mission Accomplished" on this date, in reference to the Iraq War. Note his "vivid basket" portrayed here, an obviously not missing body part. Source: Getty Images.

and the military and its weapons in the name of protection. Social and cultural beliefs about femininity inscribed onto women's bodies—softness, fragility, and nurturance—are not what make up "the few, the proud" and the hard, the shock jockeys. There exists a widespread ideology that women's bodies do not belong in the military because they are incongruent with masculinized shock treatments. In reality, though, women have been part of the military since its inception (Jeffords 1994).

In 2006, some 51% of U.S. women aged 18–24 were free to serve; 15% of active U.S. military personnel were women (U.S. Department of Defense 2006). The Air Force had the highest percentage of women (19.7%) and the Marine Corps the lowest (6%).[17] The Army had the highest percentage of black women (28%); the Marine Corps had the highest percentage of women of Hispanic origin (16%). The percentage of women officers in all branches except the Navy was similar to the percentage of enlisted personnel. In the Navy, 14.7% of officers were women in 2002, as were 13.8% of enlisted personnel (Adamshick 2005:Table 2).

Women in the U.S. military are not encouraged to serve in combat. In modern warfare, however, there isn't a clear frontline nor are there "safe" areas. In fact, the language of "noncombat zones," "demilitarized zones," and "peacekeeping missions" negates the reality that women, men, and especially children are always unsafe and at risk in these environments of conflict. In Iraq, for example, women soldiers serve as drivers in supply convoys, military police, and checkpoint patrollers, where they are used to search and question Iraqi women. In all of these jobs, these soldiers are vulnerable to roadside bombs, suicide bombers, and mortar fire, just as the Iraqi civilians living there are as well. The number of military women who have been killed and injured in Iraq is a higher percentage than in any other war since World War II, with 2% of those killed as of October 25, 2005, when the death count of U.S. military personnel reached 2,000 (Dao 2005).

Since March 19, 2003, when President Bush launched the invasion of Iraq, there have been tremendous limits placed on systems of accurate and transparent information sharing (Waldman 2004, Isikoff and Corn 2007, Rich 2006). While there have been leaks about the sexual violence perpetrated by U.S. troops against fellow U.S. soldiers and Iraqi civilians, we believe rape—like all sexual assault—is hugely underreported and hence minimized. The stories of many women are missing, lost, silenced, and invisible. At the same time, the troublesome bodies inhabited by these women cannot be ignored because of their obdurate and inescapable materiality—they do die (as in the case of women dying of dehydration or in combat zones) and they do speak (as in the case of Janis Karpinski and Jessica Lynch). Despite evidence that attempts are made to disappear or manipulate women's bodies in the context of militarization, they are recalcitrant and stubborn in the assertion of their subjectivity. Even when these bodies die, *they need to be explained.* The bodies themselves narrate a story of what happened, and they demand investigation. We are told only some stories of women in the military, and in particular one story has become a metanarrative: the story of the "rescue" of Jessica Lynch.

Speaking for Jessica Lynch

As feminist linguists Laura Prividera and John Howard report, "Understanding how the media represents women in the military is critically important because it is a primary vehicle for marking national identity and membership—who is covered and how in media stories reflect who is perceived as part of the national family" (2006:29). As we discovered, establishing an authoritative or "true" narrative of the Lynch story quickly leads one into politically contested terrain. Magazine articles have profiled the rescued soldier, biographies have been written, made-for-television movies have dramatized the events, and congressional testimony has been recorded (Bragg 2004). Each rendering of Lynch's story is partial and as such demonstrates an array of investments in its particular "moral of the story." We are not concerned here with promoting a singular version of what happened, or the ultimate "truth" of the matter, which we contend in the end is unattainable. Rather, we are interested in what the corpus of writing on Lynch has to say about gender, militarization, and the body and on the ways in which even sociological and cultural studies assessments of the rescue tend to erase, or to speak for, Lynch.

Lynch has self-reported that she quickly lost consciousness during the "ambush" events in Iraq and has not recovered these memories. (Women—and men—traumatized by violence frequently report loss of memory.) Clearly, this lack of conscious memory has proven useful for others who have used Lynch's story for different purposes. As philosopher Stephen Gallagher has noted, "She is an empty vessel for Americans to project their own fantasies, whether they are flag-waving patriotic, pro war, antiwar, feminist, anti-feminist, anything. She is a protean simulacrum, many copies, none of them referring back to any original"

(2007:127). What we recount here is how the "general public" has most commonly been told the story of Jessica Lynch, who really does exist despite philosophical claims.

Scholars have contributed a great deal to deconstructing media stories about Jessica Lynch. Cultural analysts Bruce Tucker and Priscilla Walton interpret Lynch's narrative as a story of white supremacy where classic filmic representations of Appalachia (the area of the country where Lynch is from) and its history, women's captivity narratives in American literature, and women's very presence in the military establish the compelling backdrop to the action: "Her body is a screen where everything is projected—American yearning for innocence through Appalachia, anti-feminism and anti-intellectualism through patriotism and 'hardscrabble roots,' traditional femininity over changing gendered norms, Christianity over Islam, blond white and female over dark brown and savage" (2006:315). Yet this analysis erases Lynch herself as an active person with decision-making skills and free will; the critique reestablishes the very erasure it purports to destabilize. In both the original media coverage and Tucker and Walton's analysis, Lynch is reduced to a disembodied innocent projected onto the field of action.

Susan Faludi's excellent feminist analysis of the cultural production of terrorism and its responses examines the gendered scripts that infuse our everyday lives. She charts the rise of a renewed masculinity and a disavowal of anything self-proclaimed to be "feminist" post-911: "Taken individually, the various impulses that surfaced after 9/11—the denigration of capable women, the magnification of manly men, the heightened call for domesticity, the search for and sanctification of helpless girls—might seem random expressions of some profound cultural derangement. But taken together, they form a coherent and inexorable whole, the cumulative elements of a national fantasy in which we are deeply invested, our elaborately constructed myth of invincibility" (2007:14).

Faludi's work traces through discourse and content analysis of media news the "breathless reenactments" of terror and the national preoccupation with producing superheroes (such as the men on American Airlines flight 93) and rescuing damsels in distress (such as the 9/11 widows). Faludi deconstructs the reports of Jessica Lynch's rape as written in journalist Rick Bragg's biography. Lynch has no recollection of being sexually assaulted, and Bragg cannot provide evidence of his claims: "Everyone of them [the doctors and nurses treating Lynch] said they had seen no evidence that she had been sexually assaulted" (Faludi 2007:192). In an interview with Diane Sawyer, Bragg defended leaving the sexual assault in the book, despite its lack of veracity. Faludi's analysis, echoing our own, is that "he had written a fairy tale, a cautionary one, in which the princess goes to war—and pays the price for not staying in the castle" (193). It is striking to us how readily the story of Lynch is told through the metaphorical power of fairy tales; not only Faludi and other scholars and reporters, but we also find the fairy-tale motif unavoidable when analyzing the gendered cultural production surrounding Jessica Lynch.

The military's technological delivery of the footage of captivity and rescue also attests to the power of the military-media nexus and their ongoing cooperation (O'Connell 2005). Additionally, the story of Lynch's victimization and recovery by the masculine military is used to demonstrate the failure of feminist activism and policies that have enabled women to join the military, as her body was proven to be "ill suited for military operations" (Holland 2006:28). It is as if her injuries and recovery are an enormous, "I told you so" to feminists who advocated for women's full participation in the military. Furthermore, American Studies scholar Stacy Takacs states, "by foregrounding Lynch's femininity, passivity, and vulnerablilty, the videos invite Americans to embrace militarized masculinity as the only logical antidote to national insecurity" (2005:307).

Indeed, the media narration so exclusively featured Lynch's whiteness and helpless (and hapless) femininity that her "fellow" 507th soldiers Army Specialist Shoshana Johnson, an African American wounded soldier, and Lori Piestewa, a Native American soldier killed in the "ambush," were effectively erased. Shoshana Johnson, the first African American female prisoner of war, was captured and held for more than three weeks and suffered gunshot wounds in both her legs. A single mother of a toddler, Johnson was from a military family and had hoped her military service could provide funds to pursue training at a culinary-arts school.

Johnson's truck rolled over, and she and another specialist hid under the truck to return fire. In an interview with Veronica Byrd (2004) of *Essence* magazine, Johnson states, "I got off one round and then my gun jammed. All of our weapons jammed because of the sand, so we had no way to return fire. Then I felt a burning sensation in my legs." A second Marine rescue on April 13, 2003, retrieved the five remaining POWs from the 507th ambush, including Johnson. *Washington Post* reporters Peter Baker and Terry Neal (2003) claim that the Marines were not convinced Johnson was an American. She states, "At first, they didn't realize I was American. They said, 'Get down, get down,' and one of them said, 'No,

Retired Army Specialist Shoshana Johnson, the first African American woman prisoner of war, pictured here at the White House Correspondents dinner on May 1, 2004. Source: Getty Images.

Vietnam veteran Archie Ortiz saluting a picture of Pfc. Lori Piestewa at a memorial on the Navajo Indian Reservation in Tuba City, Arizona, April 5, 2003. Piestewa was the first Native American woman killed in combat on foreign soil. Source: Getty Images.

she's American.'" We find this confusion notable because it points both to the chaos of engagement and to the incongruence of Johnson's racialized female body in such a setting. Not only was Johnson's nonwhite body perceived as "foreign" on the battlefield, raising disturbing questions about the sort of profiling that might go on in the heat of war, but her postwar experience was also deeply racialized.

As columnist Richard Koonce reported in November 2003, Johnson was overlooked as a media "hero" despite receiving the Bronze Star, the Purple Heart, and the Prisoner of War medal for her service in Iraq, and her service was actually given less value by the military. Lynch, discharged as a private first class, received an 80% disability benefit, whereas Johnson received a 30% disability benefit from the Army. Commenting on this difference, Rudolph Alexander, a social work professor, writes, "Shoshana Johnson, however, was literally ignored, except for meeting with the Congressional Black Caucus, an appearance on BET (Black Entertainment Television) and marshalling a parade for Black cowboys. Shoshana Johnson's treatment is typical of African Americans in the military, who routinely have been slighted" (2005:63). Johnson herself has been silent, and a book deal based on her journal was cancelled.

Private First Class Lori Piestewa, a soldier in the U.S. Army Quartermaster Corps, was the first Native American woman to die in combat on foreign soil. Also from a military family that includes veterans from the Vietnam War and World War I, Piestewa, a single mother of two, was half Hopi and half Mexican American. During the ambush, she was driving the Humvee when a tractor-trailer jackknifed in front of her, forcing the Humvee to hit a pole. It was then hit by a rocket-propelled grenade. Piestewa died in an Iraqi hospital of catastrophic injuries.

The memorializing of Piestewa, a friend of Lynch, has been a consistent theme in the public speaking and private actions of Jessica Lynch. In 2005, the ABC program *Extreme Makeover: Home Edition* selected Piestewa's parents and children to be the recipients of a 4,300-square-foot home "makeover" after being nominated to the show's producers by Lynch.[18] Lynch's daughter is named in honor of Piestewa: Dakota (meaning "friend") Ann (Piestewa's middle name). This act of honoring Piestewa's memory through Lynch's reproductive body was noted by many popular media outlets. In our view, this is not random celebrity reporting; rather, it reasserts ideological messages about what women *should* be doing with their bodies. Jessica Lynch may have been a "failed" soldier needing to be rescued by the "real" heroes, but she redeemed these failings by fulfilling her maternal destiny and by being a good friend.

Troublesome Bodies

The hypervisibility of Jessica Lynch performs cultural work around gender and the military: women are damsels in distress, and American masculinity and military prowess is our protector. At the same time, this hypervisibility erases or hides the bodies of many other military women. Lynch becomes the only feminized military story told to the exclusion of everything else, including women's risk from other soldiers in their platoons. Certain bodies aren't actually *missing* from social institutions and official narratives—for example, women *are* in the military—as much as their presence is disruptive to business as usual. Incongruent bodies skew our vision and challenge taken-for-granted notions about gender and its place. The incongruent, often female, body becomes the exceptional case and can be used as a foil to reassert the dominance of a particular social order. For example, the presence of women's and children's bodies disrupts the bounded and calculating rationality of many public institutions. Much has been written about how bodies interfere with the normal function of institutions such as the academy, the military industrial complex, and the corporation (Kanter 1977).

Feminist philosopher Margrit Shildrick (1997) finds that female bodies especially are constructed and seen as messy, unbounded, shifting, porous, flexible, and unpredictable. These troublesome bodies persist; in other words, they continue to appear in contexts that do not accommodate them but that may routinely sexualize them. Jessica Lynch inhabits a visible yet incongruent body, one perceived as a liability on the battlefield. Small, waif-like, feminine, and without balls (actually and metaphorically), Lynch peoples a story that is circulated widely to reproduce established cultural scripts about gender. Obviously, Lynch was not safe on the battlefield, despite the rhetoric of military protection. Men cannot, in fact, keep women safe and all too often are the perpetrators of violence against women. The three unnamed women who died from dehydration because they were afraid of being raped while urinating, and Shoshana Johnson, Lori Piestwa, Abeer Qassim Hamza, and hundreds of other girls and women—all are at risk from institutionalized aggression and sexual violence. Ultimately, militarization and the rhetoric of male heroism not only do not protect women (or men), but they reproduce the structural circumstances that make women vulnerable.

In the end, the fairy tale is proven to be the most popular version of the truth. The official story is that the princess was rescued and lived happily ever after. She even had a baby. Men are heroes, and (some) women are saved. But when the princess is allowed to speak for herself, as in Lynch's congressional testimony, she interrupts the fantastic configuration of "truth" and war as reality TV. The fairy tale is contradicted and its narrative derailed. In this perversion of the expected flow of information and gendered realities, we can see the many contradictions embodied by women in the military. Women in the military inhabit a tense combat zone, caught between sexualized visibility and gendered invisibility.

Next, we turn our attention to a starkly different kind of action figure, one whose body reaffirms cultural scripts about male dominance and superhuman strength. In competitive cycling and oncology, the intertwined theaters of war in which Lance Armstrong is a bankable star, American masculinity takes a few hits but ultimately emerges triumphant.

7

It Takes Balls

Lance Armstrong and the
Triumph of American Masculinity

What's the deal with that name, anyway? Lance Armstrong. Is
that a comic-book hero or a bendable action figure? Once somebody gives you a name like that, how hard can life be? Lance
Armstrong. Wasn't he the star of those 1950s boys' sports books?
LANCE ARMSTRONG, ALL-AMERICAN HERO!

Rick Reilly, "Sportsman of the Year" (2002)

Human bodies are fragmented, divided into specific parts for
unique purposes. In allopathic medicine, for example, we rarely have our
entire bodies x-rayed or examined; rather, body parts are isolated as part
of the doctrine of specific etiology, considered distinct from the organism
as a whole. We suffer headaches, stomachaches, and backaches, the pain
displaying corporeal regionalism. When we exercise at the gym, we work
particular muscles: one day our biceps and the next our quads. We know
how, thanks to fitness magazines, exercise shows, and the ever-present in-
structors and trainers. In pornography, we see only body parts shown in
exquisite detail, the full body often not making the frame. The camera
lens zooms in on penises, vaginas, and breasts, writhing, heaving, spread
open for visual consumption. Our body parts, the pieces that comprise
the human machine, each have a history. Some, such as breasts and faces,
are highly visible, while others are hidden, tucked away in our cellular
folds and blood-rich cavities. All body parts are laden with significance.
The story of our own breasts, for example, could be framed as a narrative
about girlhood, puberty, sexual florescence, pleasure and anxiety, body
image, infant feeding, aging, and health.

In this chapter, we are especially interested in meanings of male anatomy. The human scrotum, for example, is not an everyday topic of conversation. Outside of sex, pornography, and the omnipresent television image of professional athletes adjusting their cups on the playing field, testicles—the spherical glands dangling inside the scrotum—are invisible. Sometimes packaged creatively to enhance size and appearance (recall President Bush's "basket" from chapter 6), men's balls are nonetheless routinely hidden inside clothing and absent as body parts from public discourse. Pornography, when it does focus on male genitalia, tends to emphasize the phallus in all its rigid glory and not the "family jewels" gilding the sword. Diseases of men's bodies (e.g., prostate and testicular cancers) are not nearly as well known or oft discussed as those affecting women's bodies. Indeed, testicles are more popular as a metaphor denoting masculinity: having balls means being a man. Correspondingly, having no balls, like Jessica Lynch, means that one is feminized, a so-called "pussy."

Lance Armstrong, the subject of our inquiry here, has legendary balls. This is entirely appropriate, given that his name conjures up images of spears, javelins, and, dare we say it, another thrusting object: the phallus. Few professional athletes have achieved the megasuperstardom and instant name recognition of the seven-time Tour de France champion. While his achievements on a racing bike are unparalleled and the stuff of sporting legend, his identity as a testicular cancer survivor has further propelled Armstrong into the public eye. Indeed, he remarks often in media interviews that he would prefer to be known first as a survivor and second as a Tour champion. Certainly many athletes have donated their names to various causes, but no athlete has achieved the kind of commingled integration of sport, charity, and celebrity embodied—literally—by Armstrong. And few athletes have been more entrepreneurial: Lance Armstrong™ is an icon relentlessly self-fashioned—physically, mentally, and culturally.

Commenting in 2003 on an Annie Leibovitz photograph of the famous cyclist, journalist Rachel Koper writes: "When I look at Lance Armstrong's thighs I get weak in the knees. The sinuous calf, the knee straining like a neck . . . then those thighs. Naked, with tan lines, head down in the rain on the bike. . . . Since the picture was actually shot indoors and not at an actual race, the lighting is fairly even and bright, and Lance becomes a breathing emblem of toughness—an avatar of endurance. It's easy to ignore the tan lines from those goofy spandex tights and the fake rain because those thighs don't lie" (2003). The Austin, Texas, reporter is not the only person obsessed with Armstrong's ripped thighs. *Sports Illustrated*'s Rick Reilly describes a black-tie event in Las Vegas at which golfer Tiger Woods, himself no slouch in the fame and fortune department, asked Armstrong if he could feel his legs: "And Tiger took his hands and put them on Armstrong's concrete thighs. 'Man!' he said, squeezing. 'I mean, man!'" (2002:52).

As alluring as Armstrong's thighs may be—and we admit to a certain regret that we have never fondled them in the name of science—we are equally interested in what lies *between* the man's awesome quads. The story of Lance Armstrong, über-cyclist, cannot be told outside of the story of Lance Armstrong, testicular cancer survivor. And in media accounts, autobiographies, biographies, and Lance Armstrong Foundation materials, this is exactly how the legend is narrated. Both these battles—to overcome advanced cancer and to become a champion athlete—have intertwined to make Armstrong into a mythic (and lucrative) figure.[1] In some ways, it's a familiar trope: an ordinary young man in his prime is struck down by cancer, undergoes aggressive treatment, recovers miraculously, and is a changed man, better than he ever was as a person and an athlete. As the cyclist himself frequently declares, "The truth is that cancer was the best thing that ever happened to me."[2]

But Armstrong's illness narrative, just like Lynch's rescue tale, is no simple fable: it is complicated and fueled by his celebrity, which begins and ends with that legendary, much-photographed physique. Whether vulnerable in disease or triumphant in victory, his body and its extraordinary visibility in popular culture have contributed to the making of the man and the myth. Journalist Martin Dugard (2005) describes "chasing Lance": following Armstrong and the other cyclists around France, with the man in yellow typically pedaling furiously at the front of the *peloton*. It seems to us that everyone has been chasing Lance Armstrong: rival athletes, photographers, attractive women and men, curious and awestruck children, sponsors, journalists, cancer survivors, scientists, and a couple of sociologists. What can one famous body and its highly visible machinations tell us about masculinity, illness, sports, philanthropy, and the redemption of American national identity in wartime?

Testicular Cancer and the Politics of Men's Health

Testicular cancer is the most common type of cancer in young American men ages 15 to 34. It is highly treatable, compared with many other cancers, if diagnosed early. The disease is characterized by development of a malignancy (or malignancies) in the testicles, which are located inside the scrota underneath the penis. The testes produce sex hormones and sperm cells for reproduction. Causes of testicular cancer are unknown, although there is some evidence that it may be linked to environmental toxins (Daniels 2006). For example, the U.S. Centers for Disease Control and Prevention (CDC) suggests a connection between pregnant women's use of diethylstilbesterol (DES) and development of testicular cancer in male offspring.[3] The age-adjusted incidence rate in the United States during 2000–2004 was 5.3 per 100,000, and the median age at death for testicular cancer was 40 years. Unlike many other diseases where incidence and mortality are higher in African Americans (as with infant mortality rates), for testicular cancer the incidence among white men is significantly higher, at 6.3 per 100,000 for whites as compared with 1.4 per 100,000 for African Americans.[4]

Men's health issues, especially those concerning the genitals, historically have been invisible, and only recently has a men's health movement emerged to rival the women's health movement (Clatterbaugh 2000, White 2002). As sociologists Dana Rosenfeld and Christopher Faircloth (2006) suggest, studying medicalization, men's bodies, and men's health can enhance our understanding of masculinity and gender relations. For example, in *Sperm Counts*, Lisa chronicled practices involving sperm in medicine and culture, including interpretations of semen and male genitalia, noting a clear connection between sperm and constructions of masculinity. All too often, serious discussions of male anatomy (despite visualization in pornography, for example) are shrouded in secrecy, denial, and shame. Men do not often talk about their health, seek care on their own, or undertake preventive practices such as testicular self-exam (TSE). For example, our classroom suggestions to male students that they should be practicing such exams have been routinely met with discomfort and embarrassment. This general neglect of men's health "generates considerable pain and suffering, along with sizeable and avoidable health care costs" (Meyer 2003:709).

The consequences of testicular cancer may be significant, even when the disease is not fatal. In a qualitative study of men's most humiliating experiences, the rankings were, from the most to the least, as follows: not maintaining an erection during sex, *losing a testicle to cancer*, being teased about penis size, having a rectal exam, being diagnosed as sterile, being left by an intimate partner, and being seen naked by male friends (emphasis added).

Canadian psychologist Maria Gurevich and her colleagues assert "testicular cancers occur at a point in a man's life when the impact on sexuality, identity and fertility may be significant." Drawing on earlier studies suggesting important links between "testicular integrity" and "the coherence of male (sexual) identity," the authors interviewed 40 men diagnosed with testicular cancer. They found that the loss of a testicle was interpreted as a challenge to masculinity; the anatomical structure served as an important marker of identity. In their words, "the routes to readings of masculinity inevitably pass through anatomy . . . [and] anatomically intact bodies are designated as anatomically and socio-culturally 'correct' bodies" (Gurevich et al. 2004:1604).

In a comparative analysis of breast, testicular, and prostate cancer, medical sociologist Juanne Clarke found that, in media portrayals, "the threat of the disease seems to be less a threat to life itself than a threat to the proper, i.e., gendered enactment of life" (2004:549). Coverage of testicular cancer, in particular, emphasized early detection and aggressive treatment while also couching genital terminology in colloquialisms such as "nuts," "balls," and "family jewels." Testicles were frequently associated with manhood and masculinity, and the disease and its treatment often discussed in militarized metaphors. Reference is made in this article to Armstrong's autobiography, in which he describes "the war on cancer" and the disease as "just like a big race." Moreover, in Clarke's study, testicular cancer was found to be associated with sexuality, fertility, and relationships with women. This type of media coverage causes Clarke to ask, "Why are breast, testicular, and prostate cancers portrayed as threats to masculinity, femininity and sexuality rather than as mechanical and organic failures that could have life-threatening consequences?" (549).

Journalist Arthur Allen (1999) reports "advances in chemotherapy and other treatment nearly assure survival for most of the patients diagnosed with testicular cancer nowadays, a fact obscured by Armstrong—and most of the press—when they proclaimed his accomplishment as downright miraculous." According to oncologist Bruce Roth, however, "it is absolutely not a miracle" (quoted in Allen 1999). Testicular cancer is quite survivable, even when it is advanced. But "this is not to say that [it] is a walk in the park," writes Allen (1999). Orchiectomy, or amputation of the testicle(s), is one common treatment for testicular cancer, and potentially the one with the most lasting emotional consequences. Chemotherapy and radiation are also key weapons and may reconfigure the body while destroying malignancies. The overall message of testicular cancer is that it is often silent, rarely deadly, but can have enormous implications for a man's sense of masculinity.

"Two Lance Armstrongs": The Making of a Champion and a Cause

In 1996, when Armstrong was 25 years old, his strong, young body betrayed him by developing cancer. He did not pay attention when his right testicle swelled and became painful. In fact, he assumed it was a bike-related injury and, like many men, ignored it. In his autobiography, *It's Not about the Bike*, he writes, "Of *course* I should have known that something was wrong with me. But athletes, especially cyclists, are in the business of denial. You deny all the aches and pains because you have to in order to finish the race" (Armstrong and Jenkins 2000:5). He continued to compete, winning the Flèche-Wallonne (the first American to do so) and the Tour Du Pont. He was frequently exhausted but told himself to "suck it up." That year, he dropped out of the Tour de France after just five days, too tired and sore to be a viable contender. In September, he experienced a "brain-crushing" headache, and one day soon after he began to cough up blood. He thought perhaps his sinuses were acting up. It was not until his right testicle had expanded to the size of an orange that he sought medical care (testicles are typically about the size of a plum). The diagnosis: stage 3 testicular cancer, which had already metastasized to his lungs, abdomen, and brain.

Armstrong describes his illness as "humbling and starkly revealing," forcing him to consider aspects of his life with "an unforgiving eye" (Armstrong and Jenkins 2000:4). Indeed, the narrative of *It's Not about the Bike* follows Armstrong's cancer experience from shocking diagnosis to incredible recovery, with a brief sojourn through his personal history, including his strong bond with his mother, Linda, who raised him on her own. The subtext of the book, co-written with prolific sports journalist Sally Jenkins, is about how a champion is forged from adversity, both on and off the bike. In Armstrong's case, that adversity comes in the form of a personal and public war against testicular cancer.

The story is suffused with elements of masculinity, from Armstrong's characteristically male denial of his illness, to his relationship with his now ex-wife ("Don't be a skirt," he tells her when she drives too cautiously), to his abundant and renowned cycling achievements. The take-home message is that "there are two Lance Armstrongs, pre-cancer and post"; both are men, but only one can become a superhero. Ironically, Armstrong had to lose a testicle to gain the kind of symbolic balls that turned an ordinary, if highly successful, athlete into a megastar. Dugard puts it this way: "The cancer had reshaped Armstrong's body, stripping away all that upper body

Cover of book. Lance Armstrong and Sally Jenkins, *It's Not about the Bike: My Journey Back to Life* (Penguin), 2001.

musculature. In its place was the stick-thin torso the world has come to know so well. His heart, by contrast, was bigger—not physically, but metaphysically. Lance Armstrong had faced death and miraculously returned to life. He knew what it was to race against time" (2005:81).

In Armstrong's own story, cancer is deeply transformative: "I left my house on October 2, 1996, as one person and came home another. I was a world-class athlete with a mansion on a riverbank, keys to a Porsche, and a self-made fortune in the bank. I was one of the top riders in the world and my career was moving along a perfect arc of success. I returned a different person, literally. In a way, the old me did die, and I was given a second life" (Armstrong and Jenkins 2000:4). One of the most striking aspects of Armstrong's account is the fear and self-doubt that consumed him after he received his cancer diagnosis. Before cancer, he believed himself to be "an indestructible 25-year-old, bulletproof" (14). He worried not only that cancer might take his career and his life but also that it would change his very definition of *self*. He writes, "There were gallons of sweat all over every trophy and dollar I had ever earned, and now what would I do? Who would I be if I wasn't Lance Armstrong, world-class cyclist?" Immediately post-diagnosis, he was reduced to a frighteningly monolithic identity: "sick person" (14).

Armstrong sought and received aggressive treatment for his advanced cancer, from October through December 1996 at Indiana University Medical Center. He underwent two surgeries: one to remove his affected testicle, and the other to remove cancerous tissue from his brain. In his autobiography, Armstrong writes: "I spent the first weekend on the couch recovering from the surgery. The anesthesia made me woozy, and the incision was excruciating. I rested and watched football while my mother cooked for me, and we both read up on cancer, exhaustively" (Armstrong and Jenkins 2000:84). He experienced two rounds of chemotherapy and later was hailed by pharmaceutical manufacturers as a poster child for their products.[5] These chemical cocktails also came to have a starring role in the doping scandals and allegations by the French media of Armstrong's use of performance-enhancing supplements. "I had no life other than chemo," Armstrong recalls; "My old forms of keeping dates and time fell by the wayside, substituted by treatment regimens" (132).

An important part of Armstrong's cancer narrative is confronting his low sperm count and future reproductive capacity. He eventually banked his sperm, and after treatment, he fathered three children with whom he is frequently photographed. Only after he retired from competitive cycling did he divorce their mother and begin a succession of relationships with celebrity women. A significant portion of *It's Not about the Bike* and its sequel, *Every Second Counts* (2003), is devoted to Armstrong's quest for his lost masculinity. This process began with his successful efforts, with then-wife Kik, to produce children through in vitro fertilization, but it certainly did not end there. Armstrong's life, as detailed in these books, is devoted to hard bodily work, pain and suffering through sport, the annihilation of his opponents in the Tour, and a series of risky practices including diving headfirst off a 50-foot bluff into Dead Man's Hole near his home in Texas, just to remind himself that he's still alive. His is an epic quest, framed in the language of conquest. Or, as journalist Daniel Coyle (2006) terms it, *Lance Armstrong's War: One Man's Battle against Fate, Fame, Love, Death, Scandal, and a Few Other Rivals on the Road to the Tour de France*.

Armstrong was not the first celebrity to struggle against testicular cancer, and probably he will not be the last. In 1970, the Chicago Bears' Brian Piccolo died at age 26 from the disease, inspiring a book and made-for-TV movie, *Brian's Song*. As sportscaster Bob Costas has pointed out, viewing *Brian's Song* is practically a male rite of passage with requisite emotional catharsis: "There's no question that Brian Piccolo's story was amplified by the movie. And now generations later, you don't know how many guys who ordinarily would be loath to admit that they shed a tear, will tell you at the drop of a hat, I still cry every time I see *Brian's Song*."[6] Emotional displays aside, the movie does not mention testicular cancer or make any overt reference to the type of cancer that Piccolo died from.

In 2000, comedian Tom Green turned his testicular cancer into television entertainment, offering real-time coverage of his surgery and filming several public service announcements. And Olympic figure skater Scott Hamilton underwent treatment in 2003 for advanced testicular cancer, accompanied by significant media exposure. Yet "it wasn't until Lance Armstrong . . . won the Tour de France [in 1999] that the disease again received mass attention" (Vastag 1999). This time around, it was an unprecedented amount, fueled by our collective fascination with sport and disease, and by increasing public attention to men's health (White 2002). Health educator Samantha King argues that Armstrong's "very public battle with testicular cancer has helped strengthen the profile of men's cancers in general" (2006:xvii). Or, as Allen (1999) puts it, "Skater Scott Hamilton, subway shooter Bernard Goetz, Alexander Solzhenitsyn and the panda Hsing-Hsing all survived testicular cancer, but none stepped forward as role models. Armstrong has gripped the role with gusto."

Like the family and friends of football player Piccolo, who established the Brian Piccolo Foundation for testicular cancer research, Armstrong set up his own organization, the Lance Armstrong Foundation (LAF). He did so during his treatment for the disease and before he knew if he would recover. Consider the self-promotional language used on the LAF website: "This marked the beginning of Lance's life as an advocate for people living with cancer and a world representative for the cancer community."[7] The LAF supports research on testicular cancer, particularly the aftereffects of treatment, and it has helped solidify Armstrong's role as an authority in the war on cancer. In 2005, it awarded 27 grants to 21 institutions across the United States, totaling more than $5 million. It also provides resources for support and care of cancer sufferers; for example, the afflicted (and their families) can download or order materials to help them guide and organize treatment and recovery. In 2005, more than 28,000 survivorship notebooks were distributed.[8]

Clearly, Armstrong's "balls" have been incredibly productive. He is not just a survivor and a winner but is a champion *for* people with cancer. He transformed his illness experience—and his considerable earnings—into an effective, financially sound advocacy organization. How effective can be measured, in part, by the unexpected success of the yellow silicon LiveStrong bracelets worn by cancer survivors, athletes, youth, gymrats, cycling fans, soccer moms, presidential candidates, and health care professionals around the country. The bracelets hit the market in 2004 at various outlets (Niketown, Foot Locker, etc.), selling for $1 each. With corporate sponsorship from Nike, 100% of the proceeds from the first 5 million bracelets went directly to the LAF. The bracelets sold out before that summer's Tour de France had ended—illustrating an emergent and profitable alliance between corporations and philanthropies.[9] Why so successful? *New York Times* consumer journalist Rob Walker (2004) suggests "there's nothing even vaguely controversial or political or even provocative about a visible declaration of concern about cancer. Perhaps more crucial, the item is associated not just with a cause, but also with a heroic athlete at the peak of his popularity."

Celebrity cases draw attention to disease, and they may help spawn social movements and change. But as physician and author Barron Lerner (2006) points out, the ways in which illnesses affect celebrities may be quite different than for ordinary people. For example, while Armstrong may fit the demographic for testicular cancer incidence, he is quite unusual in other respects. For one, he survived a rare form of advanced cancer that might have killed a "lesser" human, somebody not wealthy, privileged, and extremely fit. Anthropologist and bioethicist Barbara Koenig (2001) argues, "Armstrong's dramatic recovery is atypical. . . . Over 1,500 Americans die each day from cancer. No commercials trumpet their needs or remind us that once 'saved' from cancer we will necessarily die of something else. . . . It's the miracles that sell, repeating the heroic narratives of success we are so fond of."

In Armstrong's epic recovery, our insatiable desire for evidence of medicine's achievements collides seamlessly with our need for everyday miracles and invincible heroes. But what makes Armstrong so special, aside from his considerable fame and fortune? How is it possible that he actually *survived* advanced cancer? Is he, indeed, superhuman? How was he able to redeem his masculinity despite amputation of one of his testicles? In addressing these questions, we turn next to bodily obsessions, or the ongoing cultural examination of "what makes Lance tick" (Leopold 2005).

"The Lance Armstrong Effect": Scrutinizing the Unbeatable Body

In *It's Not about the Bike,* peppered among descriptions of grueling cancer treatment and the challenges of recovery, Armstrong details the surprisingly positive changes wrought upon his body by the disease. He writes, "There was one unforeseen benefit of cancer: it had completely reshaped my body. I now had a much sparer build. In old pictures, I looked like a football player with my thick neck and big upper body. . . . Now I was almost gaunt, and the result was a lightness I'd never felt on the bike before. I was leaner in body and more balanced in spirit" (2000:224). Ironically, testicular cancer and its physical aftermath transformed Armstrong from a decent athlete into a superstar, in part by chemically resculpting his body, paring it down to perfectly meet the demands of competitive cycling. Media coverage routinely notes the cancer as a signal moment in which Armstrong shifted from merely racing to becoming a champion, marking corporeal changes as part of the legend. Dugard declares, "Before the cancer, Lance had been just another bike racer" (2005:207).

Reilly asserts, "Among professional athletes Armstrong is mythic" (2002:52). Opinion seems mixed as to whether Armstrong's success is due to one of three factors: enhancement, genetics, or sheer will. Science writer Gina Kolata (2005) argues in the *New York Times,* "The urban legends about Lance Armstrong have been circulating for years: He's superhuman, a genetic freak, the one person on the planet so perfectly made to ride the Tour de France that competitors don't have a chance." Others have described the cyclist as Herculean, evoking mythical strength. Yet we are not invested here in resolving these disagreements; we do not particularly care morally what makes the Lance machine run like the Energizer Bunny or whether he in fact used performance-enhancing drugs. Rather, we want to focus on what public discussions about Armstrong's famous physique can tell us about sport, embodiment, and masculinity. We have detected a certain obsession with the cyclist's body on the part of sportswriters, scientists, cyclists, and the media. They all want to know one thing: how a man who survived stage 3 cancer became one of the greatest athletes of all time.

Few athletes are as frequently measured and evaluated as Armstrong. Indeed, the detailed analysis of Armstrong's body represents a whole new order of fragmentation. Journalist Michael Specter (2002) writes, "Every ounce of fat, bone, and muscle on Armstrong's body is regularly inventoried, analyzed, and accounted for. I asked him if he felt it was necessary to endure the daily prodding and poking required to provide all this information, and to adhere so rigidly to his training schedules. 'Depends on whether you want to win,' he replied. 'I do.'" This scrutiny is most evident in the doping controversy surrounding Armstrong. Coyle observes, "As the world's premier cyclist, he was treated with unparalleled levels of suspicion. Armstrong was tested thirty, forty times a year, both in competition and out. . . . In 2001–02 the Postal team had been the subject of a twenty-one-month French judicial inquiry that was eventually dropped for lack of evidence" (2005: 184).

Tested randomly in the United States, France, and elsewhere over a period of many years, the cyclist has always turned up clean. Moreover, he has repeatedly and publicly denied drug use. In a clever television ad for Nike produced in 2000, he stated "Everybody wants to know what I'm on. What am I on? I'm on my bike, busting my ass six hours a day. What are you on?" (Specter 2002). In *Every Second Counts*, Armstrong avowed, "I wanted all the tests, because I knew they would come back pure. They were my only means of vindication" (2003:80). But the accusations continue to fly, especially from the French who he perpetually beat (seven times) on their own turf (Dugard 2005, "Lance Armstrong Denies Doping Report" 2006). Surely, no "ordinary" human being could accomplish what Armstrong did without enhancement, according to this camp.

Cover of book. Brad Kearns, *How Lance Does It: Put the Success Formula of a Champion into Everything You Do* (McGraw-Hill), 2006.

Other commentators, outraged by allegations of doping, believe that Armstrong is a singularly phenomenal natural athlete endowed with superior qualities. In *How Lance Does It*, for example, his longtime friend Brad Kearns (2007) blends self-help and hagiography to position Armstrong as "genetically superior." Specter (2002) describes Armstrong's body as "specially constructed for cycling," noting that "his thigh bones are unusually long . . . which permits him to apply just the right amount of torque to the pedals." He also points out that "Armstrong's heart is almost a third larger than that of an average man." (And here we assume he means physically, not metaphysically.) Coyle remarked about Armstrong in one interview, "He's the proof that Darwinism works."[10] Here, evolutionary theory is marshaled in the service of celebrity.

A group of scientists were so impressed by the cyclist's stunning ability to recover from metastatic testicular cancer that they coined the term "the Lance Armstrong Effect" to describe such "astounding therapeutic success" (Coffey et al. 2006:445). The subtext of the article, which examines cellular and molecular factors related to Armstrong's survival, is: If Lance can do it, so perhaps can other patients diagnosed with testicular cancer (assuming they have some of the same physical attributes as Armstrong). This is a theme repeated often in LAF materials designed to inspire other cancer survivors. But as Armstrong himself notes, "Basically, I can endure more physical stress than most people can, and I don't get as tired while I'm doing it. So I figure maybe that helped me live. I was lucky—I was born with an above-average capacity for breathing" (Armstrong and Jenkins 2000:4).

A third category of Lance-chasers eschews genetic explanations while also dismissing allegations of doping. This camp believes there is an obvious explanation for Armstrong's success: he works harder and "trains more than his competitors" (Specter 2002). Reilly says, "It's not Armstrong's body that wins Tours. It's his will" (2002:52). Armstrong has described his bike as his office: "It's my job. . . . I love it, and I wouldn't ride if I didn't. But it's incredibly hard work, full of sacrifices. And you have to be able to go out there every single day" (quoted in Specter 2002). The cyclist is known for his fixation on every detail about diet (including how much each morsel weighs); watts burned while riding; his weight, speed, aerodynamics; his heart rate at rest and in motion; lactic acid levels; aerobic and anaerobic abilities; and other performance-related minutiae. His workouts are legendary and have been featured in books for the masses (e.g., *The Lance Armstrong Performance Program*). On Armstrong's success, kinesiologist Ed Coyle reports: "As it turns out, it wasn't drugs or any other artificial enhancement—it was just a simple matter of determination, natural-born physical gifts and a training ethic that 99 percent of us don't come anywhere close to having" (quoted in Randall 2006).

All of these perspectives on "how Lance does it" share two important features: they agree that the champ was (and perhaps still is) relentlessly unbeatable, and they focus on the cyclist's legendary body. We are not suggesting here that Armstrong is the only American athlete ever to be so dissected by press and pundits. Others, such as Michael Jordan and Barry Bonds, have had their share of the media spotlight, including emphasis on extraordinary physical capabilities. However, Armstrong's body represents a unique conglomeration of factors. His experiences with testicular cancer and his identity as a survivor are layered inextricably with his unprecedented success as a seven-time Tour de France champion and his celebrity. His body, and his testicles specifically, represent male vulnerability at the same time that they embody the superhuman. He is one of us, but crucially not. Not only is his body perceived to be superior to the bodies of most athletes (much less vis-à-vis ordinary people), but so is his drive to win and succeed. Regardless of which perspective one adopts regarding what makes Lance tick, all are focused on the same goal: explaining his post-cancer success and articulating a unified theory about his extraordinary physical capabilities.

Consider Discovery Education's lesson plans for grades 9–12, "The Science of Lance Armstrong." (The cyclist was initially sponsored by the U.S. Postal Service, and later by the Discovery Channel.) Objectives of the lesson plans include learning that science is essential in athletic training, researching examples of sports science, explaining how new technologies have changed sports, and discovering careers in sports science. Modules of the plan focus on physiology, equipment, psychology, and training or strategy. Armstrong's body is the material used to illustrate each module. For example, under physiology the plan reads: "Heart can pump more blood per minute and beat more times than the average heart," "He was 20 pounds lighter after cancer but with the same strength," and "Has very high lung efficiency and aerobic capacity."[11] So while the lesson plan is geared toward explaining various aspects of training and strategy, contained within it are assumptions about Armstrong's corporeal uniqueness and superiority. He becomes that anomaly which sports science must *explain*. High school students can learn about drafting, drag, and lactic acid, but they will also learn that Lance Armstrong is special. There is nobody quite like him.

Armstrong's body, then, is both literally and figuratively taken apart and displayed by educators, medical professionals, the media, the LAF, high school students, and his supporters and rivals. The physical vulnerability showcased by his testicular cancer is reframed, over and over again, as an *opportunity*. In the end, Armstrong still has only one ball (his friend, the comedian Robin Williams, jokingly calls him the UniBaller [quoted in Reilly 2002:52]), but he has *balls*, and thus his masculinity is redeemed. The reality of Armstrong's body as "deformed" (i.e., missing a testicle) is erased; the most gendered physical consequence of testicular cancer—amputation of the balls—is hidden from view. We suspect that a woman athlete who undergoes mastectomy would not be able to overcome public scrutiny and be seen as whole, nor would she forever be identified positively (much less humorously) as the one-breasted champion. Yet we see Lance Armstrong's body as whole, and then some. He is not reduced to a missing testicle but becomes even *more than* the sum of his parts—magnificent thighs, giant heart, outsized lungs, and imposing will.

These dynamics are not only gendered, but they are also thoroughly racialized. Consider the figure of Barry Bonds, the bad boy of baseball. Unlike Armstrong, who has deflected allegations of steroid use while lesser men (and women) have fallen, Bonds has been unable to dodge accusations of chemical enhancement. And where Armstrong embodies the straight, white, male American hero that everyone (except the French) admires, Bonds is unavoidably black. He is routinely portrayed in the media as a liar and a criminal, and much is made of his arrogance and lack of polish. As author and critic John Ridley (2007) points out, Bonds is just not that "likeable"; little wonder that a biography of the famous hitter is called *Love Me, Hate Me: Barry Bonds and the Making of an Antihero* (Pearlman 2007). Bloggers have picked up on the distinctions between the cyclist and the baseball giant, noting that Armstrong is often portrayed as the "anti-Barry Bonds." Not only is Bonds already culturally suspect by virtue of his race, he did not overcome a near-fatal disease nor has he been seen bicycling among the wildflowers with Dubya. His philanthropy, the Bonds Family Foundation, helps underprivileged youth (read: African American) in the San Francisco Bay Area, not cancer survivors. And the lived realities of poor kids' lives cannot compare with the glamour of celebrity diseases and charities.

Lance Armstrong is, to borrow Rick Reilly's words, a bendable action figure. He is a superman for our times. And if "it's not about the bike," then it really must be about the balls or, rather, the structural arrangements of gender that are routinely advantageous to men, particularly white men, and disadvantageous to women and men of color. As Christi Anderson of Eurosport tells Dugard in *Chasing Lance:* "It's all about having a big set. . . . The upper ten percent of the riders all have a big set of balls. Lance only has one, but it's a very big one" (2005:173). And in our enduring national quest for heroes, bigger really is better.

Holding Out for a (National) Hero

In 2006, the U.S. Navy announced that its elite SEAL (sea, air, and land) unit would begin recruiting from within the ranks of accomplished athletes, including snow boarders, surfers, ice climbers, and water polo players (Hsu 2007). Due to conflicts in Afghanistan and Iraq, in which the SEALs are heavily involved, there is an ongoing need for new recruits who can successfully complete the rigorous training. The SEALs are known as one of the toughest and fittest military units in the country, and "the Navy Special Warfare Command figures that ultra-athletes have the physical and mental toughness to get through SEAL training's legendary Hell Week and thrive in the secretive, intensely demanding world of special ops" (Liewer 2006). Because less than 25% of each class of trainees graduates, the Navy has become more creative in how it seeks and retains recruits. This means targeting men who "are already living an athletic lifestyle" (Liewer 2006), men just like Lance Armstrong who would make the perfect Navy SEAL (he was once a competitive swimmer)—except that he is now several years past the cutoff age (28 years old in 2008) for new recruits.

Armstrong's physique and its visibility evoke a dream of the perfect soldier: a lean, mean fighting machine, capable of withstanding extraordinary challenges. Armstrong is a kind of weaponized cyborg, ready to be deployed down (or up) a mountain—his memoirs are full of military metaphors involving conquests and battles. His body has been (re)crafted from a potent mix of chemicals, exercise, training, strategy, supplements, measurements of all kinds, food, and a perfect melding of his muscular frame to the technology of a super light bicycle. No surprise that "he was the first cyclist in the Tour to use aerodynamically tapered handlebars for the final time trial" (Specter 2002). Indeed, according to Specter, one of the things the French do not like much about Armstrong is that he embodies a "technological renaissance" that troubles many traditionalists who believe the Tour should be won by sweat alone. Armstrong

the cyborg is more like the soldier of the future, genetically and technically enhanced, than he is like the Tour de France riders of old (e.g.,
Thompson 2006).

A bionic man for the 21st century, Armstrong first won the Tour in
1999, *after* recovering from testicular cancer; his seventh and final victory came in 2005. In March 2003, President Bush launched an invasion
of Iraq, and just a few short weeks later, Pfc. Jessica Lynch was "rescued"
by U.S. troops. Later that month, in April 2003, Secretary of State Colin
Powell warned France that there would be consequences of its refusal to
support military action against Iraq. According to BBC News, "the idea
would be to send a signal to the French that relations are in the freezer.
President Chirac should not expect an invitation to the White House"
("Q&A" 2003). French goods and services—including wine, cheese, and
hotels—faced a backlash from American consumers who supported the
invasion (Bhatnagar 2003). Foods such as French fries were temporarily renamed Freedom Fries, a move that the U.S. Congress supported.
In such tense times, it was terribly satisfying for many Americans when
Armstrong beat France on its own geopolitical terrain.

As sportswriter Andrew Hood (2003) reported, "What could be more
galling to the Gauls on July 27 than to see Lance Armstrong—whose re-
cord, cocksure manner and red-white-and-blue, government-sponsored
U.S. Postal Service team screams American domination—atop the po-
dium on the Champs-Elysées for a historic fifth straight time? After all,
Armstrong has not just dominated their Tour the last four years, he has
changed its character by introducing private jets, bodyguards, and retract-
able barriers to a sport whose charm was once rooted in the accessibil-
ity of its champions." It mattered little that Armstrong publicly stated his
opposition to the invasion of Iraq, especially after he was photographed
bicycling with his friend George W. Bush on his Texas ranch.

Like Jessica Lynch, who left the war with significant injuries requiring
technical intervention, Armstrong, too, has become a pop culture cyber-
netic organism. Feminist historian of science Donna Haraway's (1991) cyborg was meant to be gender-free, a kind of explosion of binary categories
such as male and female and an opportunity for hopefulness about the
dismantling of gender through technological integration and innovation.
Yet we want to argue here that cyborgs are profoundly gendered; they may
deepen gender differences and cleavages rather than erase them. Jessica
Lynch and Lance Armstrong both are cyborgs, to be sure, but they are
very different kinds of cyborgs: one a girl soldier, configured as a fragile,

injured princess to be rescued, the embodiment of "messy hardware," and the other a superhero for our times, a soldier of the future (and of fortune), remade stronger than ever before.

In closing, we want to talk about what happens when visually privileged superheroes become human again. In 2006, Armstrong took on a different challenge: he ran the New York City Marathon, a grueling 26.2-mile race through Manhattan. By his own admission, he had not really prepared, other than engaging in his usual postretirement workouts, and he stumbled across the finish line exhausted, sweaty, and obviously in pain. He finished the race in the middle of the pack, with a time of two hours, 59 minutes and 36 seconds—"good stockbroker or schoolteacher time" (Cazeneuve 2006:23). Signaled out among 37,000 runners by his celebrity, a "Lance Cam" captured his efforts for a live webcast. One marathoner said, "I don't know how Lance's race went, but I almost hope it was pretty tough on him, because that gives us more credibility" (23). There was grumbling among serious racers about the marathon being "hijacked" by celebrities; one remarked, "There were two American Olympic medalists in this race and you hardly knew it. If Madonna wants to run next year, what are you going to do?" (quoted in Jeansonne 2006). Armstrong himself remarked upon finishing, "The two-hour guys in front. I don't know how they do it" (Cazeneuve 2006:23)

But even with his now "imperfect" body clad in baggy shorts and his less-than-championship finish, Armstrong is redeemed as a genuine hero. Perhaps not as strong as he was in his Tour days, to at least one commentator, he is still a winner: "So when he crossed the finish line yesterday on a perfect Technicolor fall day . . . it was not only a victory for a first-time marathon runner, but for thousands of cancer patients like me and survivors everywhere. . . . I applaud him every day for what he has done for the thousands of cancer patients and survivors worldwide who look to him for daily inspiration. The fact is, people with cancer run a marathon every day, whether it's going to work or walking a block to the subway or doing laps around the hospital ward after a grueling surgery. So if he inspires us to get out and walk or move or live, he deserves a lot of credit" (Terrazzano 2006).

In victory or defeat, with just one actual testicle but a collectively recognized set of symbolic cojones, Lance Armstrong is a mythical hero for our time, in ways that Jessica Lynch, with her delicate femininity and "failure" as a soldier, could never be. Okay, so the über-jock did not win the New York City Marathon, nor is he likely to be recruited by the SEALs for Operation Iraqi Freedom. But his elevation to national superstardom, both in the cycling world and in the war on cancer, is predicated on his technically enhanced, exuberantly displayed, visually choreographed masculinity. He is a "man's man," a frequent subject in magazines catering to men's interests and the very picture of the valiant warrior from children's fairy tales. After beating cancer, Armstrong wrote, "I want to die at a hundred years old with an American flag on my back and the star of Texas on my helmet, after screaming down an Alpine descent on a bicycle at 75 miles per hour" (Armstrong and Jenkins 2000:1). This sure sounds to us like the embodiment of glorious, red-blooded, American masculinity, reproduced and celebrated in the name of national allegiance.

8

Conclusion

Excavations

It was an awful time, 2007, what Queen Elizabeth II might have called an *annus horribilis*. The Iraq War was in its fourth bloody, immoral year. In April, a deeply disturbed young man executed 32 people at Virginia Polytechnic Institute and State University and wounded many others before killing himself. A fiery helicopter crash took four lives and burned hundreds of acres in the mountains east of Seattle. Girls and women were raped, beaten, and murdered. A bridge collapsed in Minneapolis, smashing and trapping victims underwater. Desired embryos were germinated and miscarried, and thousands of infants died. Angry fires raged across Southern California. And in Afghanistan, a young American medic was killed in combat. This was by no means the only, or even the most significant, death caused by U.S. military engagements in the world's battle zones. But it was the most intensely personal for us: the Marine was Monica's brother-in-law, Luke Milam.

A Special Amphibious Reconnaissance Corpsman assigned to Golf Company of the 2nd Marine Special Operations Battalion, Luke was on his fourth combat tour. He had previously served three tours in Iraq. On September 25, 2007, the fighting in Afghanistan's Helmand Province was intense, with many casualties on all sides. Luke's unit was heavily engaged, and his position on top of a Humvee exposed him to a direct attack. Fatally injured by recoilless rifle fire, Luke bled to death within minutes after being hit. As noted by all who eulogized him in Seattle and Denver after his body was returned home for commemoration and burial, Luke died "doing what he loved." We note here the cruel irony that Luke decided to enlist in the Navy as a medic—essentially a healer of warriors—after losing his best friend, Isaiah Shoels, to the gun violence at Columbine High School in 1999. Wars of all kinds inevitably bring death, and families of soldiers and civilians alike grieve and mourn.

We have long been aware of the dangers posed by the world around us and by our precarious existence on the planet, even more so since we became mothers to small creatures. People live, and they die. Human beings harm one another, irrevocably. We know the exquisite pain of living and the losses it brings, but knowledge does not necessarily bring acceptance. Indeed, one of our aims in this book is to challenge complacency about disease and death—including the kind of social death produced through ignorance, callousness, and brutal inequality. Immersed in this project for the past several years, and in the morbid curiosities chronicled here, we often felt as if our own corporeal borders had become increasingly porous. As if the pain, suffering, grief, and debilitating harm we investigated had entered our lives through our skin and our senses, shaping our subjectivities in profound ways. With our colleagues in trauma studies (e.g., Kaplan 2005, Miller and Tougaw 2002), we believe this is more than empathy and is rather something akin to but not quite *affect;* it is a kind of embodied sociological opening to the mess and horror of the world (Cho 2008).

As sociologists of health and illness, our research projects have always engaged the body: its anatomy, threats to it, and its fragility and obduracy when confronted by disease, organized medicine's power, new technological imperatives, and the hazards of living. Medical sociology provides a framework—or, rather, a set of frameworks—within which to situate the suffering, morbidity, and mortality that punctuate (and puncture) everyday life. We use sociological tools to understand the social and political worlds within which bodies circulate and gain meanings, in particular offering a critical analysis of the hegemony of allopathic health care systems and biomedical models (Clarke et al. 2003). Whether the topic is fetal surgery, commodification of sperm, sterilization technologies, genital anatomies, or any of the subjects discussed in this book, we have found that there are no easy answers to "the problem of the body" and its location in and across biopolitical worlds. Drawing on a range of analytical strategies, we have attempted here to chart the uses and misuses of human bodies, creation of various subjectivities, and operations of power in numerous settings. Our hope is that this project intervenes not only in the literature on biopower but also our "home" field: sociology of health and medicine.

As we worked on this book, we were fully cognizant that had we written it before 9/11, it would have been a radically different animal. That said, we wanted to resist viewing issues of bodies and embodiment solely through the inexorable lens of terror, and indeed became annoyed every time 9/11 insinuated itself into our analysis. Yet we nonetheless were compelled by our data to return again and again to the emergence of the 21st-century security apparatus. Obviously, we are not suggesting that before the attacks on the World Trade Center and the outrageous U.S. military response, bodies were not at risk. Historically, little girls and boys were sexually assaulted and criminals prosecuted, babies died before their first birthday, HIV/AIDS devastated bodies and communities, synthetic toxins worked their way into organic beings, soldiers died or were rescued, and athletes became famous and made ridiculously large sums of money. We are arguing, however, that something *is* different now; we are living under what we term "a new ocular ethic." We insist that bodies are made invisible or rendered hypervisible in novel ways, via innovative technologies, and for sundry and often more insidious purposes. At the risk of sounding paranoid, we all, some of us more obviously than others, live under the watchful eye of a state: some states are all-powerful, others are putrifying from the inside out, and still others are being destroyed from the outside in by geopolitical enemies. But our embodied subjectivities are intimately linked to and shaped by our geopolitical locations within nation-states.

Even in this postmodern age of virtual surveillance and disembodied subjects, the theme of modernity and state formation resurfaced repeatedly and ironically in our investigation. Michel Foucault's intellectual project was to investigate the human sciences of "man," those discourses and practices of knowledge that enabled humans to know themselves beginning in the 19th century. While we certainly expected to find continuing evidence supporting Foucault's claims—after all, we study biopolitics—we were intrigued with how often the bodies, objects, systems, technologies, knowledges, politics, diseases, people, and states we studied were made thoroughly modern through their own actions and through actions upon them. Infant mortality, it turns out, is not merely a pre-modern problem nor is it limited to pre-modern "others" in the developing world; it is fundamentally caught up with the transnational operations of modernity, including institutionalized racism and demography. HIV/AIDS is not just a 20th-century pandemic; it represents the intersection of viral invaders with human hosts in the broader context of nations in various "states" of hardiness and decline. The diseased subject produces the failed state produces the abject Other; denigrating the humanity of "the foreigner" serves to make "us" (especially in the West) more modern.

Both biopolitics and necropolitics are activated in the ongoing quest for Modernity. Yet something about the framing of these ideas by Foucault,

Achille Mbembe, and others suggests that unitary subjectivities are produced: that disciplinary power and biopower together shape a certain kind of subject—*the* patient, *the* mad individual, *the* criminal. Each of these operates at the individual level, as a kind of Weberian ideal type, and also at the level of populations as an aggregate category. Also, this framing suggests that necropolitics shapes a specific embodied response to power; as Mbembe (2003) argues, subjectivities are formed out of the response of imperiled individuals to the nefarious operations of nation-states. While we do not disagree with these formulations—and, indeed, have relied on and critically engaged them in this book—we argue that there is no one single subject produced through discursive and institutional machinations. Even within the category of "the patient" there are multiple configurations of the patient stratified by race, gender, class, and citizenship status. Among "the mad" there are wide variations in experiences and representations of mental illness. And criminality is highly racialized in the United States particularly, with certain bodies (African American and Latino) far more likely to be incarcerated than other (white) bodies. Our analysis has attempted to show that through the operations of biopolitics and necropolitics, varied subjectivities are produced, and these have much to do with already extant inequities and social locations.

We promised earlier to offer a theory of visible and invisible bodies. The first thing we need to do is state that some of the same conditions and properties lead to both visibility and invisibility but *in thoroughly stratified ways*. Race, for example, fosters the social concealment of some infant deaths but amplifies the perceived risks posed by black men on the down low. Age is often but not always connected to status. Babies, children, the elderly (especially women), and the sick have little agency and power, while most white men, especially those in the Pentagon, White House, and other halls of political influence, have a surplus. Gender, class, sexuality, disability status, citizenship status, and geography: all are social locations with robust material consequences that shape the workings and effects of visibility and invisibility. Thus, visibility and invisibility are stratified in manifold ways. Bodies—seen and hidden, lost and found, alive and dead, actual and virtual—bear the marks of power and the many local and global processes through which it produces subjects.

We spoke earlier of an ocular ethic. We are offering conceptual tools here for how social scientists and others might learn to see the invisible, to track the ghosts embedded in our social and material lives. We want to suggest that how we see is as important as the degrees and operations of visibility and invisibility. It matters not only what we look at but also how we look. Because we are proposing engaged scholarship about subjects that are difficult to perceive, or are institutionally "protected" and therefore inaccessible, as sociologists we believe that creative ethnographies are required. Methodological "inventions," if you will, must be employed. (There is clearly a reason why Piaget researched his own children.) Limited access to subjects is gained through alternate routes by using secondary data sources such as children's books, blogs, and textual and discursive media. But due to the predominance of sciences of the aggregate— epidemiology, demography, criminology—individual stories and bodies are often lost. We have tried to make use of available aggregates, but in a critical way, to see what is done to subjects and bodies through the channeling of their flesh and bone into rates and percentages, policies and programs. Our method requires continued persistence, pushing back at the institutions and statistical methods that deny us entrée and that work to hide actual suffering.

The most useful way for us to further elaborate our theoretical points is to revisit the themes of this book: innocence, exposure, and heroism. Each chapter in this book charted some of the conditions under which certain bodies become visible or remain hidden. We propose, then, that the dimensions of visibility depend, to some degree, on what category bodies fit into. Are the bodies in question innocent, exposed, or heroic? Of course, these are not mutually exclusive categories (nor are they the only categories "out there" that may help us frame these issues). Bodies traffic between and among the categories, and each category is associated with varied benefits and liabilities. Indeed, the categories themselves, it seems, are highly reliant on one another. For example, to be innocent, you need limited exposure. To be a hero, you must have a lot of exposure. And to overcome exposure, you must be able to make claims about your innocence, and potentially emerge as a victorious hero.

As bodies are able to move between innocent, exposed, and heroic, different degrees of visibility emerge and attach to actual persons. Now we see them, now we don't; now we care, now we won't. Importantly, there is significant cultural work required to maintain bodies in the realm of innocence for as long as possible. It is exposure of "the innocent" to ideas, things, and bad people that renders their original innocence suspect. For the most part, we attempt to avoid exposure for our babies, children, selves, citizens, and humans in general. Self-protection is an imperative. Yet exposures occur, and risks to potential threats are hypervisibilized.

Diseases strike, wars are instigated by despots, babies die, soldiers and civilians are maimed and killed by explosive devices. If exposures can be prevented or overcome in dramatic ways, stories can be retold as heroic. Lance Armstrong overcomes testicular cancer and becomes a hero; he is accused of doping and proclaims his innocence. Jessica Lynch returns from war injured but washed clean of the horrors of capture by her celebrity and patriotism—at least until media exposure reveals her "questionable" allegiance and re-pathologizes her.

Innocents are kept innocent by their invisibility. Their ghostliness, managed by social forces ranging from the family to media, renders their unseen-ness a prerequisite for their innocence. An interactive relationship exists between, on the one hand, being contained in a state of being unseen and, on the other hand, being spoken *for*. Typically, the innocents remain powerless to narrate their own stories. The physical and "speaking-self" body is missing from the discourses of infant mortality and child predation—rather, these figures are fabricated by and in culture and presented to us for rational (adult) consumption. Babies and children are unable to represent themselves, especially if they are victimized or dead. This is not, of course, the same thing as denying they have agency.

Yet the bodies of the innocent also retain some quality of being seen, or hinted at, in the cultural imaginary. Infant mortality as a demographic category, especially as it is configured through prevention campaigns such as preconception care, disciplines new parents to be afraid of child death—and parents do become afraid, even though only rarely (for the most part) seeing dead babies. *To Catch a Predator* never actually depicts the sexualized bodies of children, but an imaginary is required to grasp the horror (or perhaps the allure) of these fresh bodies being abused, particularly when the fantastic acts are described in seductive terms by the show's narrator or provocative transcripts are read from the sting operation.

Protecting the fragile and unseen status of the innocent, such as through biomonitoring, is a way to reproduce the notion that they are untainted by corrupting forces that may mark them or, worse, kill them. So they are held in a perpetual state of concealment, but always with the potential threat that they will be harmed. And so the adults responsible for them, and other children socialized by the narratives of these "at risk peers," at the same time manage the invisibility of babies and young children and are disciplined by the unseen bodies in need of protection.

For example, as the press and subsequent commentary around the 2008 documentary film *Please Talk to Kids about AIDS* suggests, fierce parental protection obscures children's access to information. This point of view is given voice by Kristina from Nevada City, California, who writes to the *New York Times* coverage of the film, "There is no appropriate way to teach young children about AIDS and we should not try. . . . It is time to stop robbing children of their innocence in this society by forcing adult concepts and images on them at a far too young age."[1]

Varying degrees of visibility and the ways we are kept from seeing certain young bodies encourage and persuade the general public to fabricate stories about these bodies and then to invest confidence in other institutions, such as the criminal justice system or demography. We then see some bodies only as part of an aggregate. For the most part, innocents are invisible. When visible, they possess ideal types of bodies: consider the white, blue-eyed Gerber Baby or the pure, nonsexual child constructed in the American cultural imaginary. Rational disciplines emerge ostensibly in the service of those made invisible; yet, at the same time, these bodies of knowledge contribute to concealing the bodies further or only making them visible in extremely limited ways, such as through numbers or transcripts that are blackened out to protect the "innocent" (minor) whose case is dissected on national television. These partial glimpses shape how we perceive the innocent and guide our yearning for innocence.

Exposure can erase innocence. At the very least, exposure can mar or invalidate claims to innocence. Fear of exposure is in the ether: perhaps if we hide these bodies and make them invisible, then the threats cannot get to them. But there are unintended consequences of maintaining children's innocence and limiting their exposure. As in Ian Fleming's *Chitty Chitty Bang Bang,* hiding the children to protect them from the evil childcatcher who smells their bodies results in the children living in a dungeon: underground, fearful, and sickly and prey to other dangers lurking in these dark spaces.

But how, paradoxically, are exposed bodies also kept from view? Hidden in plain sight, if you will. It used to be that the bodies of people with AIDS were highly visible. Karposi's syndrome bruised disfigured faces, and wasting syndrome decimated bodies. In previous wars, the harmful effects of weapons on bodies were not only seen but reproduced. Recall the memorable and widely circulated photograph of Kum Phúc, the Vietnamese girl running down a road near Trang Bang after being napalmed by U.S. troops. Where are the bodies of war now? Where are the victims of exposure to various munitions? What benefits accrue to the state and to global capital from hiding these injured bodies?

While we might not see bodies exposed to HIV/AIDS or pollutants, we do hold an image of them in our heads as something unknown but also imagined that makes us afraid. We live in a state of fear of exposure as an invisible threat to bodies that needs to be managed by security forces, whose presence also exaggerates the threat. Unable to prevent certain exposures ourselves, we turn to the biosecurity apparatus to manage these for us. But we then must have faith that such protectors have our best interests in mind, or we must remain ignorant of how power works. Subjectivity in such settings is confused and confusing. Where is agency in the quest for protection? How do we both resist and succumb to the promise of safety?

The War on Terror, Bioshield, human biomonitoring, devices for cleansing the blood: biopower is at play here, but these technologies also require extraordinary leaps of faith that they work and are safe. Exposures are measured and mitigated for us, but at the same time, the fear of the security apparatus may do scary things to our bodies. What are the acceptable risks we take without even consenting by relying on the security apparatus? Does fear make us less innocent? Do we become hardened and skeptical, suspicious and cynical, and thus less than pure? How does exposure change us? Can we ever get back to the state of innocence? If not, perhaps this is what puts such a premium on innocence in that it is always forever lost.

What work does concealment of exposed bodies do? It maintains a constant state of fear without evidence, for one. We cower in the face of aggregate threats posed by nondistinct bodies, as if that teeming, writhing mass of people with AIDS is not really composed of individual suffering bodies with feelings and blood, lives and relationships. Rather, these bodies assume a monstrous quality that threatens us and works against their need for compassion and care. We are safe, but they are not. There are ethical dimensions to revealing these relationships, as well as being willing to see them.

Heroes are highly visible bodies, and their visibility prevents us from seeing alternative bodies that might also be deemed heroic in other circumstances. There are superheroes among us, victors and warriors, and we seem to crave them. We are rarely moved by the everyday hero—for example, the single mom who goes to work, comes home, feeds her kids, and keeps it all together. Our superheroes are made visible by the media, corporations, the military, sporting event sponsors, Congress, the president, and of course, heroes themselves. They are visible to all of us through digital circuits of information and capital. Their hypervisibility teaches us the power of celebrity, how to be a patriot, what it means to be a "clean" athlete, how we might go about being "rescued," what practices enable us to be properly gendered, and above all, how to be healthy.

A hero also requires a degree of innocence: you cannot have a guilty hero (or even one who is guilty until proven innocent by the circumstances of the crime). Yet heroism also demands degrees of exposure to risk and threat. If there is no threat, then there is no basis for the heroic action. Ostensibly, you cannot be awarded a Purple Heart without putting your body in the line of fire (although sometimes these precious items are given out erroneously and gratuitously, as when they are awarded for political reasons). When our heroes do fall, they fall hard. Consider the consequences if Lance Armstrong is ever definitively proven to have engaged in doping; might he suffer the same humiliation and imprisonment as former Olympic-winning athlete Marion Jones?

What happens when we excavate the missing, the sick, the dead, the destitute, and the lost? Try, if you are brave enough, to see children as sexual agents, or as able to narrate their own stories for themselves. What if children (and their bodies) were not pathologized, if they had recognized desires of their own and were not depicted exclusively as victims of sexual predators? Why is predation seemingly the only discourse we have about threats to children's bodies? And why are abstinence and criminalization considered the only acceptable responses to these hypersexualized threats?

What if we saw dead babies every day when these losses happened? What if newspapers carried obituaries for dead infants, or we enabled some register other than that of demography for making sense of infant mortality? What if we could see people with AIDS as human beings, as individual suffering bodies and not as global security threats? Framed as a public health issue, there might be less money for treatment, to be sure, but there might also be more humanity and compassion.

What if the bodies of heroes were less visible, and the bodies of the abject more visible? Without the superhero, who or what would we believe in? Without the superinfector or the terrorist, who or what would we fear?

In our framing, the ocular ethic represents the responsibility that comes with seeing or perceiving bodies *and* identifying and recovering those bodies that are unseen or less exposed to public view. It is at the same time a presumptuous and political act with weighty consequences to engage in this enterprise. This book could not have been written without a delicate intellectual balancing act—keeping the invisible and visible in view simultaneously. Drawing on our multitasking skills (the special province of working mothers, it seems), we pushed ourselves to comprehend the deeply interactive and iterative processes by which the seen and unseen are constructed and the costs and benefits of (trans)national attention. While we examined bodies that are disproportionately represented, especially in the United States, we also explored how their representation eclipsed other bodies—those that are missing—from full consideration.

Using our emergent ocular ethic—so necessary and vital in this age of surveillance and security—we focused our attention on excavating hidden, concealed, and buried bodies and on revealing the found objects of our investigatory practices. In this respect, our work is kin to sociologist Nikolas Rose's "somatic ethics" (2007). But in order to consider these found objects—such as children's agency or a baby's health—we needed to dissect and display the structural and symbolic processes that led to their social erasure in the first place. The ocular ethic means that we must confront the cultural imaginary that creates a titillating multimedia spectacle of children's sexualized bodies or the national disregard for its tiniest, most vulnerable citizens.

The ocular ethic then transforms us as we probe beneath the surface, to the ghosts lurking inside and haunting structures and practices, to the bodies caught up and made invisible in rational discourses. Our own innocence is profoundly shaken, and we are scarred by the continuous revelations of inequity, misplaced compassion, and blind nationalism in the service of creating ever more fascinating examples of the innocent and the heroic at the expense of the exposed. The dead women who could not urinate for fear of sexual assault, the multigenerational families affected by HIV/AIDS, the babies buried in silence, the fallen soldiers, the abject sick cast as threats to our stability and order, the victims of chemical poisoning of all kinds including state-sanctioned corporatism—these bodies haunt us still.

But these missing bodies also motivate us to see and do more. The 21st-century security guards tell us that if we see something, we should say something. We believe that an ocular ethic needs to appropriate this strategy in subversive ways. We have indeed seen something—bodies concealed and revealed through the transnational and localized operations of biopower—and so we have said something, hopefully significant, in the pages of this book. We hope that readers may be similarly provoked to turn the oppressive slogans of contemporary surveillance on their heads, and instead to work at revealing the mechanisms of state power as embodied in millennial practices and subjects.

Notes

CHAPTER 1

1. There is considerable high-speed traffic in pornography, fashion, sports, diseases, communication technologies, photography, cinema, organs, sex work, labor, migration, medical tourism, war, and other practices.

2. At http://www.nytimes.com/2007/01/17/world/middleeast/17iraq. html?_r=1&oref=slogin.

3. In an interview with www.takegreatphotos.com, photographer Vincent Laforet commented, "VL: The best example of that is the one I took of a woman on a conveyor belt [while covering Hurricane Katrina]. I felt sick to my stomach even making that photograph, but you bring yourself to do it because it's a responsibility at that point. And in fact the person from FEMA there begged me to take those pictures. She said, 'We're overwhelmed. We have 900 people an hour coming in. We can't deal with it. I need more people here. You need to take these pictures.' Literally. The FEMA person. You would think it would be in total opposite interest for me to take those pictures, but it wasn't about ego; it's wasn't about careers anymore; it was just about saving these people and helping them. And that transcended all the bureaucratic b.s. that we deal with all the time. And when you make that photograph, it suddenly becomes a burden because you know you have to get it back to the paper. Making it is just half the story; you have to somehow get it back. So we had to race back in the helicopter to Baton Rouge to get a cell signal to transmit just in time for the deadline. There was a lot of pressure. A lot, a lot of pressure to make sure you got it."

CHAPTER 2

1. "The Danger of Child Sexuality," Foucault's dialogue with Guy Hocquenghem and Jean Danet, was produced by Roger Pillaudin and broadcast by France Culture on April 4, 1978. It was published as "La Loi de la pudeur" in *Recherches 37*, April 1979. First published in English in *Semiotext(e) Magazine* (New York): Semiotext(e) Special Intervention Series 2: Loving Boys / Loving Children (Summer 1980), in a translation by Daniel Moshenberg. A full version was translated by Alan Sheridan with the title "Sexuality Morality and the Law," published

in Michel Foucault, *Politics, Philosophy, Culture: Interviews and Other Writings,* ed. Lawrence D. Kritzman (New York: Routledge, 1988).

2. In the United States, throughout the history of production of childrearing expertise, multiple and often conflicting groups claim the truth about parent-child relationships. Often experts—whether medical, legal, mental health, or administrative—present themselves as more knowledgeable and valid than mothers. See, for example, Jean O'Malley Halley (2007), *Boundaries of Touch: Parenting and Adult-Child Intimacy* (Urbana: University of Illinois Press). In addition, while it may appear obvious that women are still the target of social anxieties about childrearing, men, strangers, and queers are also important suspect figures that loom in our cultural imaginary.

3. At http://www.unicef.org/crc/.

4. At http://www.unicef.org/crc/index_30229.html.

5. Also, Monica has completed a memoir titled "True Confessions of Bambi's Mom: On Trauma, Hope, and Species Survival." She is currently shopping for a publisher.

6. See H. Edgar (1992), "Twenty Years After: The Legacy of the Tuskegee Syphilis Study," *Hastings Center Report* 22, no. 6, 29–40. Includes articles by Arthur L. Caplan, Harold Edgar, Patricia A. King, and James H. Jones.

7. The Belmont Report laid out three general ethical principles that should govern human subjects research: beneficence—to maximize the benefits of science, humanity and research participants and to avoid or minimize risk or harm; respect—to protect the autonomy and privacy of rights of participants; and justice—to ensure the fair distribution among persons and groups of the costs and benefits of research.

8. At http://www.aaup.org/AAUP/comm/rep/A/humansubs.htm.

9. For more detailed analysis of children as a population for social scientific research, see Kodish (2005) and Grodin and Glantz (2006).

10. Once published, Levine's book, the author, and the University of Minnesota Press encountered considerable controversy. As reported by Minnesota Public Radio, "Conservative commentators called *Harmful to Minors* evil. And before the book was in stores, House Majority Leader and Republican gubernatorial candidate Tim Pawlenty called the book 'trash,' and tried to get the University of Minnesota Press to stop publication" (at http://news.minnesota.publicradio.org/features/200208/13_helmsm_harmful/). The National Coalition for the Protection of Children and Families' website states about Levine's book: "The most alarming aspect is the fact that the notion of legitimizing adult/child sexual relations has moved out of the 'fringe' and into mainstream academia, the same path that the sexual revolution and homosexual movements followed" (at http://www.national-coalition.org/culture/articles/levineresponse.html).

11. In another example of inflammatory accusations used to neutralize or disable sexuality discussions with children, Irvine found that sex education as a field

is likened to sexual abuse of children: "Sex education, they charged, is sexual abuse. . . . A more developed argument appeared in the widely disseminated sex education critiques by psychiatrist Melvin Anchell, who pronounced that 'seduction is not limited to actual molestation. A child can be seduced, in psychoanalytic terms, by overexposure to sexual activities, including sex courses in the classroom.' One text announced that sex education is not education at all, but rather a 'legalized form of child seduction and molestation' (2004:133).

12. "But managing children is not limited to what they do with their bodies and when. Curricula and classroom practices, themselves socially constructed and labile, exert control over what, and how, a child thinks in relation to his or her own and others' bodies. And nowhere is this more apparent than in relation to sexual curiosity" (Granger 2007:4).

13. "Pedophiles see themselves as part of a social movement to gain acceptance of their attractions" (Eichenwald 2006). For pedophilia in the criminal justice context of what is being erased, see this source.

14. Statistics strongly suggest that sexual abuse occurs overwhelmingly by someone who is known to the victim. For example, Bureau of Justice statistics from 2000 reveal that sex offenders of children under 6 were strangers in just 3% of the cases and 5% in cases involving 6–11 year olds (U.S. Department of Justice 2000). Shows like *To Catch a Predator* actually eliminate familiar assailants as an option, instead emphasizing stranger cases.

15. Perverted Justice recruits volunteer contributors who pose as underage children in chatrooms. Posing from a variety of ages (standard ages are 10–15), these contributors simply go into chatrooms with fake online screen names and wait for predators to instigate conversation with them. See http://www.pervertedjustice.com/. Beginning in the fourth season of *TCAP*, revelations of fees paid to Perverted Justice led to criticism of journalistic conflict of interest.

CHAPTER 3

1. *March of the Penguins*, official website, at http://wip.warnerbros.com/marchofthepenguins/.

2. At www.rottentomatoes.com (accessed October 30, 2007).

3. At http://www.who.int/child-adolescent-health/OVERVIEW/HNI/neonatal.htm.

4. At https://www.cia.gov/library/publications/the-world-factbook.

5. Historian Jan Doolittle Wilson argues that the bill, campaigned for by the Women's Joint Congressional Committee, reflected Progressive-era interest in protecting children and searching for "legislative solutions to child-related problems" (2007:29). Following on the heels of the 1918 National Children's Year Campaign, the bill enjoyed wide support from both major parties in Congress, President Wilson, the Children's Bureau, and other agencies.

6. Race is complicated, and racial classifications are reflective of social structure. We draw on Michael Omi and Howard Winant's (1986) definition of racialization as "the extension of racial meaning to a previously unclassified relationship, social practice, or group. Racialization is an ideological process, an historically specific one."

7. The infant mortality rate of Native Hawaiians and Pacific Islanders is more
than 31% greater than that of whites, while the rate for American Indians and
Alaska Natives at 8.1% is almost twice the rate for whites. Asians have a lower
infant mortality rate than whites but the highest rate of infant deaths from birth
defects for reasons that remain unclear. Overall, Hispanics do not have higher
mortality than other groups, and among certain ethnicities such as Mexicans, the
rate is actually lower than the U.S. national rate, suggesting that infant mortality
is complicated by both race and ethnicity.

8. The Thomases argued that if people define situations as real, they are real
in their consequences.

9. Monica's broader project on the biopolitics of infant mortality will include
an analysis of the emergence of PCC. See also Fordyce (2008).

10. At http://www.marchofdimes.com/pnhec/173_14005.asp.

11. At http://www.mymagicalmemories.org/.

12. At http://home.mend.org/.

13. Formerly at http://www.sidsfamilies.com/letterstoheaven/letters.cg.

14. At www.firstcandle.org.

CHAPTER 4

The subtitle of this chapter is from Colin Powell, in 2004 as secretary of State (as
quoted by Greg Behrman in the *Dallas Morning News,* July 31, 2004, at http://
www.cfr.org/publication/7226/global_aids_threat.html).

1. The Millenium Challenge Corporation (MCC) was established in 2004 and
is supported by congressional allocations. For a country to receive funding from
the MCC, it must promote political and economic freedom, invest in education
and health, control corruption, and respect civil liberties. In 2008, the chair of
the board of directors was Condoleeza Rice; other members included the secretary of the Treasury, the U.S. trade representative, the administrator of USAID,
Senator Bill Frist, and various venture capitalists.

2. At http://www.usatoday.com/news/world/2008-02-16-bush-africa_N.htm.

3. At http://www.pbs.org/newshour/bb/africa/july-deco3/africa_7-7.html.

4. At http://data.unaids.org/pub/EPISlides/2007/071118_epi_revisions_
factsheet_en.pdf.

5. The CDC has also calculated estimated numbers of AIDS cases by year of
diagnosis and transmission type. The figures show that as of 2005 in the United

States, a total of 984,155 citizens have AIDS. Just over 9,000 of these are children below the age of 13 at diagnosis. Among the 761,723 men diagnosed with AIDS between 2001 and 2005, at least 452,111 (ca. 59%) contracted the disease through male-to-male sexual contact. Another 168,314 men (ca. 22%) contracted the disease through injection drug use. These are the two most prevalent causes of HIV/AIDS among men in the United States. Among women, 73,050 (ca. 40%) of the 181,802 with the disease contracted it through injection drug use. High-risk heterosexual contact accounts for the majority of cases, 102,171 (ca. 56%). The remaining cases were caused by other factors such as hemophilia, blood transfusion, perinatal exposure, or another, unknown, risk factor (CDC 2007:table 3).

6. At http://data.unaids.org/pub/EPISlides/2007/2007_epiupdate_en.pdf.

7. At http://www.nytimes.com/2008/01/06/health/06HIV.html.

8. At http://hrw.org/reports/2007/us1207/index.htm.

9. At http://www.fas.org/irp/threat/nie99-17d.htm.

10. Laurie Garrett reports, "In 2000, the U.N. Security Council issued Resolution 1308, warning that the HIV/AIDs pandemic, if unchecked, could threaten world stability and security" (2005a).

11. At http://archives.cnn.com/2000/HEALTH/AIDS/04/30/aids.threat.03/index.html.

12. At http://usinfo.state.gov/archives.old/products/washfile/latest/2002/December/02120503.clt.

13. George Tenet, Director of Central Intelligence Agency's Worldwide Threat Briefing, "The Worldwide Threat in 2003: Evolving Dangers in a Complex World," February 11, 2003, at http://www.cia.gov/cia/public_affairs/speeches/2003/dci_speech_02112003.html.

14. The *CIA World Factbook* collects and disseminates data on HIV/AIDS for every country in the world alongside data on transportation, communication, morbidity and mortality, economy, military, exports, and so on. Of particular interest, HIV/AIDS is the only disease with its own line item.

15. The United States is not the only nation to redefine AIDS as a security threat. The deputy prime minister of Russia, Alexander Zhukov, stated that "the spread of HIV/AIDS is beyond a medical problem, and . . . it has become an issue of strategic, social and economic security of the country" (World Bank 2005).

16. At http://www.nytimes.com/2008/01/05/washington/05aids.html?ex=12002 00400&en=ce22d8fc7d6fabe8&ei=5070&emc=eta1.

17. At www.pepfar.gov.

18. At http://www.nytimes.com/2008/01/20/us/20castro. html?ei=5070&en=c8b1bde4bc69173a&ex=1201582800&adxnnl=1&emc=eta1&adx nnlx=1201443099-a4+PYwBoYqdmxyLxoYkMMg.

19. At http://www.cwfa.org/main.asp.

CHAPTER 5

1. Supporters included the American Nurses Association of California, California Medical Association, California Nurses Association, California Primary Care Association, San Francisco Bay Area Physicians for Social Responsibility, San Francisco Medical Society, Breast Cancer Fund, California NOW, Endometriosis Association, DES Cancer Network, Marin Breast Cancer Watch, Women's Foundation of California, Nursing Mothers Counsel, Commonweal, California League of Conservation Voters, Center for Environmental Health, Children's Health Environmental Coalition, Communities for a Better Environment, Environment California, Environmental Justice Coalition for Water, Environmental Working Group, Greenaction for Health and Environmental Justice, Healthy Building Network, Healthy Children Organizing Project, Natural Resources Defense Council, New Leaf Paper, Pesticide Action Network, Pesticide Free Zone Campaign of Marin, Planning and Conservation League, Sierra Club California, Silicon Valley Toxics Coalition, WORKSAFE!, United Steelworkers of America, American Federation of State, County, and Municipal Employees, San Francisco Labor Council, Vote Health, Trust for America's Health, National Brain Tumor Foundation, Marin Golden Gate Learning Disabilities, Learning Disabilities Association of California, California Church Impact, California Interfaith Partnership for Children's Health and the Environment, Los Angeles Interfaith Council, California Safe Schools, California Seniors Coalition, and Latino Issues Forum. At http://www.breastcancerfund.org.

2. The California Body Burden Campaign engaged the marketing firm Lake-Snell-Perry and Associates to conduct a statewide survey on public perceptions of the relationship between toxic chemicals and disease. The survey found that California voters in every demographic group ranked cancer, obesity, heart disease, and diabetes as their top health concerns. When asked specifically about the link between pollution and illness, 97% agreed that toxic chemicals and industrial pollutants could cause health problems and disease. Over one-half of voters indicated "extreme" concern about toxic chemicals in their bodies, and over three-quarters of voters believed it was very important to know what was in their bodies. Significantly, over three-quarters expressed support for a biomonitoring program regardless of the funding mechanism, although voters had more favorable responses to the idea that companies producing toxins would be assessed a fee to support the program. At http://www.calbbc.org/site/pp.asp?c=9elELMMAG+b=65900

3. Among the coalition's members were the American Chemistry Council, American Electronics Association, American Forest and Paper Association, BIOCOM, California Chamber of Commerce, California Farm Bureau, California Independent Oil Marketers Association, California Independent Petroleum Association, California Paint Council, California Retailers Association, California

Women for Agriculture, Chemical Industry Council of California, Chlorine Chemistry Council, Crop Life America, Fresno County Farm Bureau, Grocery Manufacturers of America, IPC California Circuits Association, Rubber Manufacturers Association, Silicon Valley Leadership Group, Surface Technology Association, Soap and Detergent Association, and Western States Petroleum Association. At http://www.americanchemistry.com.

4. Letter to California State Senate, April , 2006, at www.ca/circuits.org.

5. At www.americanchemistry.com.

6. Department of Toxic Substances Control (2007) news release, May 15.

7. At http://www.dhs.ca.gov/ehlb/bpp/default.htm. See also California Department of Health Services (2003. According to sociologist Rachel Washburn, "The new California Environmental Contaminant Biomonitoring Program (CECBP) was devised by the Breast Cancer Fund and Commonweal, as per SB 1379. This bill stipulated the development of the program, and so far—from what I have gathered from several meetings—the parameters of this program have been in line with the bill. The extent to which the California Biomonitoring Plan has been integrated is unclear" (personal correspondence, April 29, 2008).

8. At the time of this writing in spring 2008, the Environmental Working Group was calling its effort the Human Toxome Project. Their Internet site stated: "Just as scientists raced to define the human genome, the Human Toxome Project (HTP) at Environmental Working Group is working to define the human toxome—the full scope of industrial pollution in humanity. Using cutting edge biomonitoring techniques, the HTP scientists, engineers, and medical doctors test blood, urine, breast milk and other human tissues for industrial chemicals that enter the human body as pollution in food, air, and water, or from exposures to ingredients in everyday consumer products" (at http://www.bodyburden.org/).

9. At http://indigenouswomen.org.

10. As human biomonitoring expands, the privileged position of breast milk may shift. According to sociologist Rachel Washburn, "Some folks I have talked with have started to move away from [breast milk biomonitoring] a bit on the grounds that there are other ways to assess infant exposures. That said, I don't think anyone is discounting it as a way to assess exposures . . . it's just that there are a number of chemicals now being measured that are not necessarily best measured in this fluid" (personal correspondence, April 7, 2008).

11. At http://www.calbbc.org.

12. The first Technical Workshop on Human Milk Surveillance and Research on Environmental Chemicals was held in February 2002 at the Milton S. Hershey Medical Center, Pennsylvania State University, College of Medicine. A follow-up workshop was held in 2004. Articles based on presentations at this workshop were subsequently published in a special issue of *Journal of Toxicology and Environmental Health, Part A*, vol. 68 (2005).

13. Both of us breastfed our daughters, and while expressing milk into a pump rather than into our children's eager mouths was relatively easy for one of us, it was virtually impossible for the other.

14. At http://www.commonweal.org.

15. American Institute for Cancer Research, 2008, at http://www.aicr.org/site/PageServer?pagename=res_report_second. The full report and the executive summary for this study can be obtained at https://secure2.convio.net/aicr/site/Ecommerce?store_id=4581&JServSessionIdr009=5s8iie6na2.app45a.

16. At http://bcaction.org/index.php?page=newsletter-75f.

17. For the full report, see http://www.fda.gov/cber/gdlns/tissdonor.pdf.

18. Thanks to Rachel Washburn for this point.

19. At http://www.cryobank.com/sbanking.cfm?page=2&sub=126.

20. As explored in *Sperm Counts* (Moore 2007), CODIS, the FBI's databank of DNA information, is made possible through a coordinated effort of federal, state, and local law enforcement agencies. Each agency collects biometric data from a variety of men and a variety of crime scenes attempting to match materials. Due to varying laws and law enforcement practices, certain types of men are more likely than others to end up with their biological information in CODIS regardless of the crime committed or proof of guilt. Due to institutional racism and inequality endemic in the criminal justice system, practices of racial profiling "capture" men of color and poor men within the biological/criminal nexus—in a mutually reinforcing cycle, these men are thus labeled dangerous and simultaneously treated as such.

21. Although this may be changing (Daniels 2006).

CHAPTER 6

1. Allegations that the rescue itself was orchestrated for dramatic impact are fairly common. The most alarming allegation may be one by Representative Henry Waxman, the oversight committee's chairman, who suggests that perhaps the military delayed Lynch's rescue by one day to allow the team to be accompanied by a cameraman. It was the resulting night-scope video that was released to the media and broadcast widely (Whitelaw 2007:39).

2. Army News Source, 2003, at http://www.militaryconnections.com/news_story.cfm?textnewsid=

3. At http://www.washingtonpost.com/ac2/wp-dyn/A14879-2003Apr2.

4. At http://edition.cnn.com/2003/US/11/07/lynch.interview/index.html.

5. At http://oversight.house.gov/documents/20080714111050.pdf.

6. In 1992, women were harassed and assaulted at the Tailhook Association convention in Las Vegas, when naval aviators "formed a gauntlet on the third floor of the Hilton and trapped women in it, pawing and molesting them, stripping off their clothes" (Douglas 2007). This event turned into an

even more impressive scandal when the Navy was revealed to have engaged in a major cover-up of the incident. Secretary of the Navy H. Lawrence Garrett was forced to resign after it became clear that he had not only approved the cover-up but also been present in one of the Tailhook party suites during the infamous convention. As feminist communications scholar Susan J. Douglas notes, "we should remember Tailhook, what it did, and did not, change in the military."

7. "Of the 12 Aberdeen staff members charged, two are serving prison terms for sexual misconduct convictions, one was cleared of sexual misconduct charges, two agreed to be discharged in lieu of court martial, four including Robinson face court martial, and three have yet to complete the military equivalent of a grand jury hearing" (at http://www.cnn.com/U.S./9705/27/army.sex/).

8. At http://hrw.org/reports/2003/iraq0703/2.htm#_Toc45709963>http://hrw.org/reports/2003/iraq0703/2.htm#_Toc45709963.

9. The quotation in the heading is from http://www.salon.com/news/feature/2007/03/07/women_in_military/.

10. After she wrote this piece, three Bosnian soldiers were indicted in 2001 for rape through the War Crimes Tribunal at The Hague (at http://www.globalpolicy.org/intljustice/tribunals/2001/0803icty.htm). Oosterveld is also advisor to the United Nations, Human Rights and Economic Law Division of Foreign Affairs.

11. At http://www.salon.com/news/feature/2007/03/07/women_in_military/.

12. At www.bushcommission.org and www.nion.us.

13. Karpinski (2005), the commanding officer at Abu Ghraib, claims she has been consistently scapegoated and targeted by the military.

14. At http://www.theportlandalliance.org/2006/july/suzanneswift.htm.

15. At http://www.democracynow.org/article.pl?sid=06/09/18/1351245.

16. Bloom died in Iraq from a pulmonary embolism, a non-combat-related ailment, on April 6, 2003.

17. At http://www.womensmemorial.org/PDFs/StatsonWIM.pdf.

18. At http://www.azcentral.com/ent/tv/articles/0413piestewa.html.

CHAPTER 7

1. Forbes estimates Armstrong's annual earnings between June 2004 and June 2005 as $28 million (at www.forbes.com).

2. At www.livestrong.org.

3. At www.cdc.gov. For an excellent account of DES, see Bell (forthcoming).

4. At www.cancer.gov.

5. The first was a combination of bleomycin, etoposide, and Platinol, and the second a combination of ifosfamide, etoposide, and Platinol.

6. Bob Costas, *SportsCentury* series, ESPN Classic, quoted in Puma (n.d.).

7. At www.livestrong.org.

8. For financial data on the foundation, see *Annual Report 2005*, at www.livestrong.org.

9. The bracelets were so popular that they spawned numerous copycats in a rainbow of colors demonstrating myriad diseases and causes and have also become fashion accessories.

10. At www.booknoise.net (accessed 2006).

11. At www.DiscoveryEducation.com.

CHAPTER 8

1. At http://www.nytimes.com.

References

"Achievements in Public Health, 1900–1999: Healthier Mothers and Babies." *Morbidity and Mortality Weekly Report* 48(38): 849–58.

Adamshick, Mark. 2005. "Social Representation in the U.S. Military Services." CIRCLE Working Paper 32, May. At http://www.civicyouth.org/PopUps/WorkingPapers/WP32Adamshik.pdf (accessed October 12, 2007).

Agamben, Giorgio. 1998. *Homo Sacer: Sovereign Power and Bare Life.* Trans. Daniel Heller-Roazen. Stanford, Calif.: Stanford University Press.

Albertini, Richard, Michael Bird, Nancy Doerrer, Larry Needham, Steven Robison, Linda Sheldon, and Harold Zenick. 2006. "The Use of Biomonitoring Data in Exposure and Human Health Risk Assessments." *Environmental Health Perspectives* 114(11): 1755–62.

Alexander, Randolph. 2005. *Racism, African Americans and Social Justice.* New York: Rowman and Littlefield.

Allan, Susan. 2006. "You Are What You Eat . . . Breathe . . . Scrub . . . Lather . . . Spray." *Ottawa Citizen,* March 5.

Allen, Arthur. 1999. "Triumph of the Cure." *Salon,* July 29. At www.salon.com.

al-Rehaief, Mohammed Odeh. 2003. *Because Each Life Is Precious: Why an Iraqi Man Came to Risk Everything for Private Jessica Lynch.* New York: Harper.

Altman, Rebecca. 2008. "Chemical Body Burden and Place-Based Struggles for Environmental Health and Justice." Ph.D. diss, Brown University.

Aquilina, Mike. 1995. "Talking to Youth about Sexuality: A Parent's Guide." *Our Sunday Visitor.* At www.osv.com.

Armstrong, David. 1986. "The Invention of Infant Mortality." *Sociology of Health and Illness* 8: 211–32.

Armstrong, Lance, and Sally Jenkins. 2000. *It's Not about the Bike: My Journey Back to Life.* New York: Putnam's.

———. 2003. *Every Second Counts.* New York: Broadway Books.

Armstrong, Lance, Chris Carmichael, and Peter Joffre Nye. 2000. *The Lance Armstrong Performance Program: Seven Weeks to the Perfect Ride.* Emmaus, Pa.: Rodale.

Associated Press. 2003. "Your Body, Your Superfund Site." *Wired,* December 28.

———. 2006. "Penguins Pack Pop-Culture Punch." December 19.

Atkinson, Kate. 2006. *One Good Turn: A Novel.* Boston: Little, Brown.

Baker, Leigh. 2002. *Protecting Your Children from Sexual Predators.* New York: St. Martin's.

Baker, Peter, and Terry M. Neal. 2003. "Seven American POWs Are Rescued: Bush: 'A Great Day for the Families.'" *Washington Post,* April 13. At http://www.washingtonpost.com/ac2/wp-dyn/A16971–2003Apr13?language=printer.

Baldwin, Peter. 2005. *Disease and Democracy: The Industrialized World Faces AIDS.* Berkeley: University of California Press.

Barstow, Anne Llewellyn, ed. 2001. *War's Dirty Secret: Rape, Prostitution, and Other Crimes against Women.* Cleveland: Pilgrim Press.

Bast, Diane Carol. 2004. "Junk Science Pervades California 'Biomonitoring' Effort." Heartland Institute, Environment and Climate News, June 1. At http://www.heartland.org/Article.cfm?artId=14987.

Beauchamp, Tom L., and James F. Childress. 1989. *Principles of Biomedical Ethics,* 3rd ed. New York: Oxford University Press.

Becker, Rick, Sarah Brozena, and Darrell Smith. 2003. "What Is Biomonitoring?" At http://www.americanchemistry.com/s-acc/bin.asp?CID-257+DID-1584+DOC=FILE.pdf.

Beecher, Henry K. 1966. "Ethics and Clinical Research." *New England Journal of Medicine* 74: 1354–60.

Behrman, Greg. 2004. *The Invisible People: How the U.S. Has Slept through the Global AIDS Pandemic, the Greatest Humanitarian Catastrophe of Our Time.* New York: Free Press.

Bell, Susan E. Forthcoming. *DES Daughters, Embodied Knowledge, and the Transformation of Women's Health Politics in the Late 20th Century.* Philadelphia: Temple University Press.

Berlin, Cheston M., Jr., et al. 2005a. "Conclusions and Recommendations of the Expert Panel: Technical Workshop on Human Milk Surveillance and Biomonitoring for Environmental Chemicals in the United States." *Journal of Toxicology and Environmental Health, Part A* 68: 1825–31.

Berlin, Cheston M., Jr., Betty L. Crase, Peter Fürst, Judy S. LaKind, Gerry Moy, Larry L. Needham, Linda C. Pugh, and Mary Rose Tully. 2005b. "Methodologic Considerations for Improving and Facilitating Human Milk Research." *Journal of Toxicology and Environmental Health, Part A* 68: 1803–24.

Bettelheim, Bruno. 1989 [1975]. *The Uses of Enchantment: The Meaning and Importance of Fairy Tales.* New York: Vintage.

Bhatnagar, Parija. 2003. "French Goods Face U.S. Backlash." CNN, March 17. At www.CNNMoney.com.

Black, Harvey. 2006. "Setting a Baseline for Biomonitoring." *Environmental Health Perspectives* 114(11): A652–54.

Blum, Linda M. 2000. *At the Breast: Ideologies of Breastfeeding and Motherhood in the Contemporary United States.* Boston: Beacon.

Bono. 2007. "Guest Editor's Letter: Message 2U." *Vanity Fair,* July, 32.

Boone, M. S. 1982. "A Socio-Medical Study of Infant Mortality among Disadvantaged Blacks." *Human Organization* 41(3): 227–36.

Boykin, Keith. 2005. *Beyond the Down Low: Sex, Lies, and Denial in Black America.* Cambridge, Mass.: De Capo Press.

Bragg, Rick. 2004. *I Am a Soldier, Too: The Jessica Lynch Story.* New York: Vintage.

Braun, Bruce. 2007. "Biopolitics and the Molecularization of Life." *Cultural Geographies* 14: 6–28.

Bristol, Nellie. 2007. "Battling HIV/AIDS: Should More Money Be Spent on Prevention?" *CQ Researcher* 17(38): 889–912. At http://library.cqpress.com/cqresearcher/cqresrre2007102600 (retrieved January 15, 2008).

Broder, John M. 2003. "A Nation at War: Prisoners of War." *New York Times,* April 2.

Brown, Lakela. 2007. "Living on the Down Low." *Essence* 37(11): 122.

Brown, Phil, and Edwin J. Mikkelson. 1997. *No Safe Place: Toxic Waste, Leukemia, and Community Action.* Berkeley: University of California Press.

Brownmiller, Susan. 1993. *Against Our Will: Men, Women and Rape.* New York: Ballantine.

Brumberg, Joan Jacobs. 1998. *The Body Project: An Intimate History of American Girls.* New York: Vintage.

Buncombe, Andrew. 2007. "Infant Mortality in Iraq Soars as Young Pay the Price for War." *Independent,* May 8.

Butler, Judith P. 1993. *Bodies That Matter: On the Discursive Limits of Sex.* New York: Routledge.

Byrd, Veronica. 2004. "Shoshana Johnson: To Hell and Back." *Essence* 166 (March): 1–4. At http://www.essence.com/essence/print/0,14882,590888,00.html.

California Department of Health Services. 2003. *California Biomonitoring Plan.* Sacramento: California Department of Health Services.

Canguilhem, Georges. 1977. *On the Normal and the Pathological.* Dordrecht: Reidel.

Caplan, Jane, and John Torpey. 2001. Introduction to *Documenting Individual Identity: The Development of State Practices in the Modern World,* ed. Jane Caplan and John Torpey (1–24). Princeton: Princeton University Press.

Casper, Monica J. 1998. *The Making of the Unborn Patient: A Social Anatomy of Fetal Surgery.* New Brunswick, N.J.: Rutgers University Press.

———. n.d. "True Confessions of Bambi's Mom: On Trauma, Hope, and Species Survival." Unpublished ms.

Cazeneuve, Brian. 2006. "Marathon Man." *Sports Illustrated,* November 13, 23.

Centers for Disease Control and Prevention (CDC). 2006. "CDC Releases National Recommendations to Improve Health of Babies and Mom." Press Release, April 20.

Centers for Disease Control and Prevention (CDC). 2007. *HIV/AIDS Surveillance Report*, vol. 17, rev. ed. At www.cdc.gov/hiv/topics/surveillance/resources/reports/2005report/ (accessed December 10, 2008).

Chad, Norman. 2005. "Good for Him, Bad for U.S." *Washington Post*, July 18, E2.

Cho, Grace M. 2008. *Haunting the Korean Diaspora: Shame, Secrecy, and the Forgotten War*. Minneapolis: University of Minnesota Press.

Clarke, Adele E. 1995. "Modernity, Postmodernity and Human Reproductive Processes c1890–1990, or 'Mommy, Where Do Cyborgs Come from Anyway?'" In *The Cyborg Handbook*, ed. Chris Hables Gray, Heidi J. Figueroa-Sarriera, and Steven Mentor (139–56) . New York: Routledge.

Clarke, Adele E., and Joan H. Fujimira, eds. 1992. *The Right Tool for the Job: At Work in Twentieth-Century Life Sciences*. Princeton: Princeton University Press.

Clarke, Adele E., and Virginia Olesen, eds. 1999. *Revisioning Women, Health and Healing: Feminist, Cultural, and Technoscience Perspectives*. New York: Routledge.

Clarke, Adele E., Jennifer Fishman, Jennifer Fosket, Laura Mamo, and Janet Shin. 2003. "Biomedicalization: Technoscientific Transformations of Health, Illness, and U.S. Biomedicine." *American Sociological Review* 68 (April): 161–94.

Clarke, Juanne Nancarrow. 2004. "A Comparison of Breast, Testicular and Prostate Cancer in Mass Print Media (1996–2001)." *Social Science and Medicine* 59(3): 541–51.

Clark-Flory, Tracy. 2007. *Iraqi Women Sell Sex for Survival*. At www.Salon.com.

Clatterbaugh, Kenneth. 2000. *Contemporary Perspectives on Masculinity: Men, Women, and Politics in Modern Society*. Boulder, Colo.: Westview.

Clough, Patricia Ticineto. 2007. Introduction to *The Affective Turn: Theorizing the Social*, ed. Patricia Ticineto Clough with Jean Halley (1–33). Durham, N.C.: Duke University Press.

Coffey, Donald S., Robert H. Getzenberg, and Theodore L. DeWeese. 2006. "Hyperthermic Biology and Cancer Therapies: A Hypothesis for the 'Lance Armstrong Effect.'" *Journal of the American Medical Association* 296(4): 445–48.

Connell, R. W., and G. W. Dowsett. 1993. *Rethinking Sex: Social Theory and Sexuality Research*. Philadelphia: Temple University Press.

CRC. 1989. *The Convention on the Rights of the Child*. New York: United Nations.

Coyle, Daniel. 2006. *Lance Armstrong's War: One Man's Battle against Fate, Fame, Love, Death, Scandal, and a Few Other Rivals on the Road to the Tour de France*. New York: Harper.

Craddock, Susan. 2004. *City of Plagues: Disease, Poverty, and Deviance in San Francisco*. Minneapolis: University of Minnesota Press.

Currah, Paisley A., and Lisa Jean Moore. 2009. "'We Don't Know Who You Are': Contesting Sex Designations on New York City Birth Certificates." *Hypatia* 24(3).

Daniels, Cynthia R. 2006. *Exposing Men: The Science and Politics of Male Repro-duction.* New York: Oxford University Press.

Dao, James. 2005. "2,000 Dead: As Iraq Tours Stretch On, a Grim Mark." *New York Times,* 26 October, A1, 14, 15.

Dicum, Gregory. 2006. "Mother Knows Best." *Grist,* November 6.

DiSano, Laura. 2006. "Measuring Up: Examining the Need to Establish State-Based Biomonitoring Programs." *Journal of Environmental Health* 69(5): 32–33.

Doan, C. 2006. "Subversive Stories and Hegemonic Tales' of Child Sexual Abuse: From Expert Legal Testimony to . . ." *International Journal of Law in Context,* at http://journals.cambridge.org.

Douglas, Susan J. 2007. "The Legacy of Tailhook." *In These Times,* May 29. At http://www.inthesetimes.com/article/3182/the_legacy_of_tailhook/.

Doyle, Jamie Mihoko, Samuel Echevarria, and W. Parker Frisbie. 2003. "Race/Ethnic-ity, Apgar, and Infant Mortality." *Population Research and Policy Review* 22: 41–64.

Dugard, Martin. 2005. *Chasing Lance: The 2005 Tour de France and Lance Arm-strong's Ride of a Lifetime.* New York: Little, Brown.

Eberstadt, Nicholas. 2002. "The Future of AIDS." *Foreign Affairs* 81(6): 22–45.

Edelman, Lee. 2004. *No Future: Queer Theory and the Death Drive.* Durham, N.C.: Duke University Press.

Eichenwald, Kurt. 2006. "Dark Corners: On the Web, Pedophiles Extend Their Reach." *New York Times,* August 21, 00–00.

Enloe, Cynthia. 2000. *Maneuvers: The International Politics of Militarizing Wom-en's Lives.* Berkeley: University of California Press.

———. 2004. *The Curious Feminist: Searching for Women in the New Age of Em-pire.* Berkeley: University of California Press.

———. 2007. *Globalization and Militarism: Feminists Make the Link.* New York: Rowman and Littlefield.

Fairchild, Amy. 2004. "Policies of Inclusion: Immigrants, Disease, Dependency, and American Immigration Policy at the Dawn and Dusk of the 20th Cen-tury." *American Journal of Public Health* 94(4): 528–39.

Faludi, Susan. 2007. *The Terror Dream: Fear and Fantasy in Post 9-11 America.* New York: Metropolitan Books.

Farhi, Paul. 2006. "*Dateline* Pedophile Sting: One More Point." *Washington Post,* April 9.

Farmer, Paul. 2005. *Pathologies of Power: Health, Human Rights, and the New War on the Poor.* Berkeley: University of California Press.

Featherstone, Mike, Mike Hepworth, and Bryan Turner. 1991. *The Body: Social Process and Cultural Theory.* London: Sage.

Feldbaum, Harley, Kelley Lee, and Preeti Patel. 2006. "The National Security Implications of AIDS/HIV." *Policy Forum.* At http://medicine.plosjournals. org/perlserv/?request=get-document&doi=10.1371/journal.pmed.0030171&ct=1 (accessed August 10, 2008).

Fenton, Suzanne E., Marian Condon, Adrienne S. Ettinger, Judy S. LaKind, Ann Mason, Melissa McDiamid, Zhengmin Qian, and Sherry G. Selevan. 2005. "Collection and Use of Exposure Data from Human Milk Biomonitoring in the United States." *Journal of Toxicology and Environmental Health, Part A* 68(20): 1691–1712.

Fine, Michelle. 1988. "Sexuality, Schooling and Adolescent Females: The Missing Discourse of Desire." *Harvard Educational Review* 58: 29–51.

Fine, Michelle, and Sara I. McClelland. 2006. "Sexuality Education and Desire: Still Missing after All These Years." *Harvard Education Review* 76(3): 297–340.

Finkelhor, David. 1981. *Sexually Victimized Children*. Glencoe, Ill.: Free Press.

Fleming, Ian. 1964. *Chitty Chitty Bang Bang*. London: Jonathan Cape.

Flynn, Kevin, and Jim Dwyer. 2004. "Falling Bodies: A 9/11 Image Etched in Pain." *New York Times*, September 10.

Ford, Chandra, Kathryn Whetten, Susan Hall, Jay Kaufman, and Angela Thrasher. 2007. "Black Sexuality, Social Construction and Research Targeting 'the Down Low' ('The DL')." *Annals of Epidemiology* 17(3): 209–16.

Fordyce, Lauren. 2008. "Birthing the Diaspora: Technologies of Risk among Haitians in South Florida." Ph.D. diss., University of Florida.

Foucault, Michel. 1977. *Discipline and Punish*. Trans. Alan Sheridan. New York: Pantheon.

———. 1978. *The History of Sexuality: An Introduction*. New York: Random House.

———. 1988. "Sexuality Morality and the Law." Trans. Alan Sheridan. In *Politics, Philosophy, Culture: Interviews and Other Writings*, ed. Lawrence D. Kritzman. New York: Routledge.

———. 2003. *"Society Must Be Defended": Lectures at the Collège de France, 1975–1976*. Trans. David Macey. New York: Picador.

———. 2007. *"Security, Territory Population": Lectures at the Collège de France, 1977–1978*. Trans. David Macey. New York: Palgrave Macmillan.

Fraser, Nancy. 1989. *Unruly Practices: Power, Discourse, and Gender in Contemporary Social Theory*. Minneapolis: University of Minnesota Press.

———. 1997. *Justice Interruptus*. New York: Routledge.

Frederick, Sharon. 2000. *Rape: Weapon of Terror*. Hackensack, N.J.: World Scientific Publishing.

Friedrich, William N., Jennifer Fisher, Daniel Broughton, Margaret Houston, and Constance R. Shafran. 1998. "Normative Sexual Behavior in Children: A Contemporary Sample." *Pediatrics* 101(4): 9.

Gallagher, Stephen. 2007. "Jessica Lynch, Simulacrum." *Peace Review: A Journal of Social Justice* 19(1): 119–28.

Garrett, Laurie. 2000. *Betrayal of Trust: The Collapse of Global Public Health*. New York: Hyperion.

———. 2005a. "The Lessons of HIV/AIDS." *Foreign Affairs* 84(4): 51–64.

———. 2005b. *HIV and National Security: Where Are the Links? A Council on Foreign Relations Report.* New York: Council on Foreign Relations.

Gatens, Moira. 1992. "Power, Bodies and Difference." In *DestablizingTheory,* ed. Michele Barrett and Anne Phillips (120–37). Cambridge: Polity Press.

Gilbert, Jen. 2007. "Risking a Relation: Sex Education and Adolescent Development." *Sex Education* 7(1): 47–61.

Giroux, Henry A. 2006. "Reading Hurricane Katrina: Race, Class, and the Biopolitics of Disposability." *College Literature* 33(3): 171–96.

Glassner, Barry. 2000. *The Culture of Fear: Why Americans Are Afraid of the Wrong Things.* New York: Basic Books.

Goldstein, Richard. 2003. "Bush's Basket: Why the President Had to Show His Balls." *Village Voice,* May 20. At http://www.villagevoice.com/news/0321,goldstein,44234,1.html.

Gordon, Avery F. 1996. *Ghostly Matters: Haunting and the Sociological Imagination.* Minneapolis: University of Minnesota Press.

Gordon, Linda. 2002. *The Moral Property of Women: A History of Birth Control Politics in America.* Urbana: University of Illinois Press.

Gostin, Lawrence, and David Fidler. 2007. "Biosecurity under the Rule of Law." *Case Western Reserve Journal of International Law* 38(3/4): 437–78. .

Granger, Colette A. 2007. "On (Not) Representing Sex in Preschool and Kindergarten: A Psychoanalytic Reflection on Orders and Hints." *Sex Education* 7(1): 1–15.

Graue, M. Elizabeth, and Daniel J. Walsh. 1998. *Studying Children in Context: Theories, Methods and Ethics.* Los Angeles: Sage.

Greenhalgh, Susan. 1996. "The Social Construction of Population Science: An Intellectual, Institutional, and Political History of Twentieth-Century Demography." *Comparative Studies in Society and History* 38(1): 26–66.

Greydanus, D. E., and B. Geller. 1980. "Masturbation: Historic Perspective." *New York State Journal of Medicine,* November.

Grodin, Michael, and Leonard Glantz, eds. 2006. *Children as Research Subjects: Science, Ethics and Law.* New York: Oxford University Press.

Grosse, Scott D., Sergey V. Sotnikov, Sheila Leatherman, and Michele Curtis. 2006. "The Business Case for Preconception Care: Methods and Issues." *Maternal and Child Health Journal* 10: S93–99.

Gross-Loh, Christine. 2004. "Don't Trash Our Bodies: Researching Breastmilk Toxins." *Mothering* 122, January/February.

Gurevich, Maria, Scott Bishop, Jo Bower, Monika Malka, and Joyce Nyhof-Young. 2004. "(Dis)Embodying Gender and Sexuality in Testicular Cancer." *Social Science and Medicine* 58: 1597–1607.

Halkitis, Perry, Leo Wilton, and Jack Drescher, eds. 2006. *Barebacking: Psychological and Public Health Approaches.* Binghamton, N.Y.: Haworth.

Halley, Jean O'Malley. 2007. *Boundaries of Touch: Parenting and Adult-Child Intimacy.* Urbana: University of Illinois Press.

Halperin, David. 2007. *What Do Gay Men Want? An Essay on Sex, Risk and Subjectivity.* Ann Arbor: University of Michigan Press.

Hansen, Chris. 2007. *To Catch a Predator: Protecting Kids from Online Enemies Already in Your Home.* New York: Dutton.

Haraway, Donna J. 1997. *Modest_Witness@Second_Millennium. FemaleMan©_ Meets_OncoMouse™: Feminism and Technoscience.* New York: Routledge.

Harris, Lynn. 2007. "Fightin' Words." *Time Out New York Kids.* 20 June. At http://www.timeout.com/newyork/kids/articles/around_town/10860/fightin-words.

Harrison, Helen, and Ann Kositsky. 1983. *The Premature Baby Book: A Parent's Guide to Coping and Caring in the First Years.* Rev. ed. Darby, Penn.: Diane Publishing.

Harvey, David. 2007. *A Brief History of Neoliberalism.* New York: Oxford University Press.

Havlik, Sheree Whitters. 2005. *Because We Love Them: Fostering a Christian Sexuality in Our Children.* Notre Dame, Ind.: Sorin Books.

Healton, C., G. Kirsch, and D. Bartelli. 1992. "Years of Potential Life Lost: Implications for U.S. Federal HIV/AIDS Funding Policy." Paper presented at the International Conference on AIDS, Amsterdam, July 19–24. International AIDS Society 8:D502 (abstract no. PoD 5680). At http://www.aegis.com/conferences/iac/1992/PoD5680.html (accessed December 10, 2008).

Heller, Zoe. 2003. "'Embedded' Reporters Are Far Too Close to the Action." *The Age,* April 1. At www.theage.com.au/articles/2003/03/31/1048962698037.html.

Higonnet, Anne. 1998. *Pictures of Innocence: The History and Crisis of Ideal Childhood.* New York: Thames and Hudson.

Holland, Shannon L. 2006. "The Dangers of Playing Dress-Up: Popular Representations of Jessica Lynch and the Controversy Regarding Women in Combat." *Quarterly Journal of Speech* 92(1): 27–50.

Hood, Andrew. 2003. "Lance de France." *Time,* June 29.

Hsu, Andrea. 2007. "Navy SEALs Seek to Build Up Their Ranks." National Public Radio, October 16.

Human Rights Watch. 2003. "Climate of Fear: Sexual Violence and Abduction of Women and Girls in Baghdad." At www.hrw.org.

Hummer, Robert A., Isaac W. Eberstein, and Charles B. Nam. 1992. "Infant Mortality Differentials among Hispanic Groups in Florida." *Social Forces* 70(4): 1055–75.

Hunter, Susan. 2003. *Black Death: AIDS in Africa.* New York: Palgrave Macmillan.

Inda, Jonathan Xavier, ed. 2005. *Targeting Immigrants: Government, Technology, and Ethics.* New York: Wiley.

Irvine, Janice. 2004. *Talk about Sex: The Battles over Sex Education in the United States.* Berkeley: University of California Press.

Isikoff, Michael, and David Corn. 2007. *Hubris: The Inside Story of Spin, Scandal and the Selling of the Iraq War*. New York: Three Rivers Press.

Jacobson, Nora. 2000. *Cleavage: Technology, Controversy, and the Ironies of the Man-Made Breast*. New Brunswick, N.J.: Rutgers University Press.

Jacquet, Luc, Jordan Roberts, and Jérôme Maison. 2006. *March of the Penguins*. Washington, D.C.: National Geographic.

Jeansonne, John. 2006. "Celebrity Watch Can Come in Cycles: With Finish Not Close, Armstrong Grabs Attention." *Newsday*, November 6, D2.

Jeffords, Susan. 1994. *Hard Bodies: Hollywood Masculinity in the Reagan Era*. New Brunswick, N.J.: Rutgers University Press.

Jenkins, Henry. 1998. Introduction to *The Children's Culture Reader*, ed. Henry Jenkins (1–37). New York: New York University Press.

Johnson, Kay A. 2006. "Public Finance Policy Strategies to Increase Access to Preconception Care." *Maternal and Child Health Journal* 10(1): 85–91.

Johnson, Kay, Samuel F. Posner, Janis Biermann, Jose F. Cordero, Hani K. Atrash, Christopher S. Parker, Sheree Boulet, and Michele G. Curtis. 2006. "Recommendations to Improve Preconception Health and Health Care—United States: A Report of the CDC-ATSDR Preconception Care Work Group and the Select Panel on Preconception Care." *Morbidity and Mortality Weekly Report* 55(RR-6): 1–23.

Jones, Alison. 2003. "The Monster in the Room: Safety, Pleasure and Early Childhood Education." *Contemporary Issues in Early Childhood* 4(3): 235–50.

———. 2004. "Risk Anxiety, Policy, and the Spectre of Sexual Abuse in Early Childhood Education." *Discourse: Studies in the Cultural Politics of Education* 25(3): 321–34.

Jones, Maggie. 2007. "How Can You Distinguish a Budding Pedophile from a Kid with Real Boundary Problems?" *New York Times Magazine*, July 22. At http://www.nytimes.com/2007/07/22/magazine/22juvenile-t.html?_r=1&oref=slogin.

Kanter, Rosabeth M. 1977. *Men and Women of the Corporation*. New York: Basic Books.

Kaplan, E. Ann. 2005. *Trauma Culture: The Politics of Terror and Loss in Media and Literature*. New Brunswick, NJ: Rutgers University Press.

Karpinski, Janis. 2005. *One Woman's Army*. New York: Miramax.

Katovsky, Bill, and Timothy Carlson. 2003. *Embedded: The Media at War in Iraq*. New York: Lyons Press.

Kearns, Brad. 2007. *How Lance Does It: Put the Success Formula of a Champion into Everything You Do*. New York: McGraw-Hill.

King, Samantha. 2006. *Pink Ribbons, Inc.: Breast Cancer and the Politics of Philanthropy*. Minneapolis: University of Minnesota Press.

Kirk, Paul L. 1953. *Crime Investigation*. New York: Interscience.

Klawiter, Maren. 2008. *The Biopolitics of Breast Cancer: Changing Cultures of Disease and Activism*. Minneapolis: University of Minnesota Press.

Klein, Naomi. 2007. *The Shock Doctrine: The Rise of Disaster Capitalism.* New York: Metropolitan Books.

Kodish, Eric. 2005. *Ethics and Research with Children: A Case-Based Approach.* New York: Oxford University Press.

Koenig, Barbara. 2001. "When the Miracles Run Out in America, Care for Dying Patients Fails to Measure Up." *San Jose Mercury News,* August 7.

Kolata, Gina. 2005. "Super, Sure, but Not More Than Human." *New York Times,* July 24.

Koonce, Richard. 2003. "Johnson Should Be Revered Like Lynch." *BGNews,* November 12. At http://media.www.bgnews.com/media/storage/paper883/news/2003/11/12/Opinion/Johnson.Should.Be.Revered.Like.Lynch-1290459.shtml.

Koper, Rachel. 2003. "Piece of Work: Lance Armstrong by Annie Leibovitz." *Austin Chronicle,* November 7.

Krause, Elizabeth L. 2001. "'Empty Cradles' and the Quiet Revolution: Demographic Discourse and Cultural Struggles of Gender, Race, and Class in Italy." *Cultural Anthropology* 16(4): 576–611.

———. 2006. "Dangerous Demographies: The Scientific Manufacture of Fear." *Corner House,* Briefing 36, July.

Laaser, Mark. 1999. *Talking to Your Kids about Sex: How to Have a Lifetime of Age-Appropriate Conversations with Your Children about Healthy Sexuality.* Colorado Springs: WaterBrook Press.

Lagnado, Lucette Matalon, and Shelia Cohn Dekel. 1992. *Children of the Flames: Dr. Josef Mengele and the Untold Story of the Twins of Auschwitz.* New York: Penguin.

LaKind, Judy S., Cheston M. Berlin Jr., and Michael N. Bates. 2005a. "Overview: Technical Workshop on Human Milk Surveillance and Biomonitoring for Environmental Chemicals in the United States." *Journal of Toxicology and Environmental Health, Part A* 68(20): 1683–89.

LaKind, Judy S., Robert L. Brent, Michael L. Dourson, San Kacew, Gideon Koren, Babasaheb Sonawane, Anita J. Tarzian, and Kathleen Uhl. 2005b. "Human Milk Biomonitoring Data: Interpretation and Risk Assessment Issues." *Journal of Toxicology and Environmental Health, Part A* 68: 1713–69.

"Lance Armstrong Denies Doping Report." 2006. *CBS News.* May 31.

Leopold, Todd. 2005. "What Makes Lance Armstrong Tick?" CNN, July 5. At www.CNN.com.

Lerner, Barron. 2006. "Famous Patients and the Lessons They Teach." *New York Times,* November 14.

Leung, Alexander, and William Lane Robson. 1993. "Childhood Masturbation." *Clinical Pediatrics* April 19: 238–41.

Levine, Judith. 2002. *Harmful to Minors: The Perils of Protecting Children from Sex.* Minneapolis: University of Minnesota Press.

Lewin, Ellen, and Virginia Olesen, eds. 1985. *Women, Health and Healing: Toward a New Perspective.* New York: Tavistock.

Lidster, C. A., and M. E. Horsburgh. 1994. "Masturbation: Beyond Myth and Taboo." *Nursing Forum* 29(3): 18–27.

Liewer, Steve. 2006. "SEALs Looking for Ultra-Athletes." *San Diego Union-Tribune*, July 29.

Lioy, Paul J., Edo Pellizzari, and David Prezant. 2006. "The World Trade Center Aftermath and Its Effects on Health: Understanding and Learning through Human-Exposure Science." *Environmental Science and Technology* 40(22): 6876–85.

Lorber, Judith, and Lisa Jean Moore. 2006. *Gendered Bodies: Feminist Perspectives.* New York: Oxford University Press.

Lyon, David. 2001. *Surveillance Society: Monitoring Everyday Life.* Maidenhead, Berkshire: Open University Press.

Marcus, George. 1995. "Ethnography in/of the World System: The Emergence of Multi-sited Ethnography." *Annual Review of Anthropology* 24: 95–117.

Mbembe, Achille. 2003. "Necropolitics." *Public Culture* 15(1): 11–40.

McClam, Erin. 2007. "Memphis Fights Its Infant Mortality Rate." Newsday.com, November 10.

Mead, George Herbert. 1934. *Mind Self and Society from the Standpoint of a Social Behaviorist.* Ed. Charles W. Morris. Chicago: University of Chicago Press.

Merleau-Ponty, Maurice. 1962. *Phenomenology of Perception.* Trans. Colin Smith. New York: Humanities Press.

Meyer, Jack A. 2003. "Improving Men's Health: Developing a Long-Term Strategy." *American Journal of Public Health* 93(5): 709–11.

Miller, Nancy K. and Tougaw, Jason (eds.). 2002. *Extremities: Trauma, Testimony, and Community.* Urbana: University of Illinois Press.

Minh-Ha, Trinh T. 1989. *Woman, Native, Other.* Bloomington: Indiana University Press.

Moore, Lisa Jean. 2007. *Sperm Counts: Overcome by Man's Most Precious Fluid.* New York: New York University Press.

Moore, Lisa Jean, and Adele E. Clarke. 2001. "The Traffic in Cyberanatomies: Sex/ Gender/ Sexuality in Local and Global Formations." *Body and Society* 7: 57–96.

Morelle, Rebecca. 2008. "Biometrics Picks Up the Penguins." *BBC News,* June 27.

Morris, Rick. 2006. *Protecting and Parenting Sexually Abused Children.* New York: Lulu Press.

Moskowitz, David, and Michael Roloff. 2007. "The Existence of a Bug Chasing Subculture." *Culture, Health and Sexuality* 9(4): 347–57.

Nardi, Bonnie A., and Yrjö Engeström. 1999. "A Web on the Wind: The Structure of Invisible Work." *Computer Supported Cooperative Work* 8: 1–8.

National Commission for the Protection of Human Subjects of Biomedical and Behavioral Research. 1979. *The Belmont Report: Ethical Principles and*

Guidelines for the Protection of Human Subjects of Research. Washington, D.C.: National Institutes of Health.

National Intelligence Council. 2000. *The Global Infectious Disease Threat and Its Implications for the United States*. NIE-99;17D. Washington, D.C.: National Intelligence Council.

Newman, George. 1906. *Infant Mortality: A Social Problem*. New York: Dutton.

Nieto, José. 2004. "Children and Adolescents as Sexual Beings: Cross-Cultural Perspectives." *Child and Adolescent Psychiatric Clinics of North America* 13: 461–77.

O'Connell, Aaron B. 2005. "Saving Private Lynch: A Hyperreal Hero in an Age of Postmodern Warfare." *War, Literature and the Arts: An International Journal of the Humanities* 17(1/2): 33–52.

Oliver, Kelly. 2007. *Women as Weapons of War: Iraq, Sex, and the Media*. New York: Columbia University Press.

Omi, Michael, and Howard Winant. 1986. *Racial Formation in the United States: From the 1960s to the 1980s*. New York: Routledge.

Ong, Aiwa, and Stephen Collier, eds. 2005. *Global Assemblages: Technology, Politics, and Ethics as Anthropological Problems*. Malden, Mass.: Blackwell.

Ong, C.-N., H.-M. Shen, and S.-E. Chia. 2002. "Biomarkers for Male Reproductive Health Hazards: Are They Available?" *Toxicology Letters* 134(1): 17–30.

Oosterveld, Valerie. 1998. "When Women Are the Spoils of War." *UN Courier*. At http://www.unesco.org/courier/1998_08/uk/ethique/txt1.htm (accessed August 10, 2008).

Parker, Christopher S., Sheree L. Boulet, and Hani K. Atrash. 2006. "Improving Women's Health for the Sake of Our Children." *Journal of Women's Health* 15(5): 475–79.

Parker-Pope, Tara. 2007. "Shhh . . . My Child Is Sleeping (in My Bed, Um, with Me)." *New York Times*, October 23. At http://www.nytimes.com/2007/10/23/health/23well.html?_r=1&ref=health&oref=slogin (accessed October 23, 2007).

Patton, Cindy. 1990. *Inventing AIDS*. New York: Routledge.

Paustenbach, Dennis, and David Galbraith. 2006. "Biomonitoring and Biomarkers: Exposure Assessment Will Never Be the Same." *Environmental Health Perspectives* 114(8): 1143–49.

Pearlman, Jeff. 2007. *Love Me, Hate Me: Barry Bonds and the Making of an Antihero*. New York: Harper Paperbacks.

Please Talk to Kids about AIDS. 2007. Directed by Brian Hennessey. Produced by Radia Daoussi and Nagila Guimaraes. At www.vineeta.org.

Plummer, Ken. 1995. *Telling Sexual Stories: Power, Change, and Social Worlds*. London: Routledge.

Pope, Harrison G. Jr., Katharine A. Phillips, and Roberto Olivardia. 2000. *The Adonis Complex: The Secret Crisis of Male Body Obsession*. New York: Free Press.

Price, James. H., Joseph A. Drake, Gregg Kirchofer, and Susan K. Tellijohann. 2003. "Elementary School Teachers' Techniques of Responding to Student Questions regarding Sexuality Issues." *Journal of School Health* 73(1): 9–14.

Prividera, Laura C., and John W. Howard III. 2006. "Masculinity, Whiteness, and the Warrior Hero: Perpetuating the Strategic Rhetoric of U.S. Nationalism and the Marginalization of Women." *Women and Language* 29(2): 29.

Proctor, Robert N., and Londa Schiebinger, eds. 2008. *Agnotology: The Making and Unmaking of Ignorance*. Palo Alto, Calif.: Stanford University Press.

Projansky, Sarah. 2001. *Watching Rape: Film and Television in Post Feminism Culture*. New York: New York University Press.

Prue, Christine E., and Katherine Lyon Daniel. 2006. "Social Marketing: Planning before Conceiving Preconception Care." *Maternal and Child Health Journal* 10: S79–84.

Puma, Mike. n.d. "Brian's Life a Song of Friendship: Courage." At http://espn. go.com/classic/biography/5/piccolo_Brian.html.

"Q&A: U.S.-French Relations on Ice." 2003. *BBC News*, April 23.

Ramazanoglu, Caroline, and Janet Holland. 2002. *Feminist Methodology: Challenges and Choices*. Thousand Oaks, Calif.: Sage.

Randall, Kay. 2006. "Man and Superman: Kinesiologist Discovers Physiological Secrets behind Lance Armstrong's Cycling Success." University of Texas at Austin. At www.utexas.edu (accessed September 24, 2007).

Rapp, Rayna. 1999. *Testing Women, Testing the Fetus: The Social Impact of Amniocentesis in America*. New York: Routledge.

Reilly, Rick. 2002. "Sportsman of the Year: Lance Armstrong." *Sports Illustrated*, December 16, 52–59.

Rich, Frank. 2006. *The Greatest Story Ever Sold: The Decline and Fall of Truth from 9/11 to Katrina*. New York: Penguin.

Ridley, John. 2007. "The Unforgivable Blackness of Barry Bonds." National Public Radio, July 23.

Riley, Nancy, and James McCarthy. 2003. *Demography in the Age of the Postmodern*. Cambridge: Cambridge University Press.

Rose, Nikolas. 2007. "Molecular Biopolitics, Somatic Ethics and the Spirit of Biocapital." *Social Theory and Health* 5: 3–29.

Rosen, Ruth. 2006. "The Hidden War on Women in Iraq." At http://www.commondreams.org/views06/0713-33.htm.

Rosenfeld, Dana, and Christopher A. Faircloth, eds. 2006. *Medicalized Masculinities*. Philadelphia: Temple University Press.

Rubin, Gayle. 1984. "Thinking Sex: Notes for a Radical Theory of the Politics of Sexuality." In *Pleasure and Danger*, ed. Carole Vance. New York: Routledge, 267–319.

Ruzek, Sheryl Burt. 1978. *The Women's Health Movement: Feminist Alternatives to Medical Control*. New York: Praeger.

Saenz, Rogelio. 2007. "The Growing Color Divide in U.S. Infant Mortality." *Population Reference Bureau,* October.

Sanderson, Christiane. 2004. *The Seduction of Children: Empowering Parents and Teachers to Protect Children from Child Sexual Abuse.* London: Jessica Kingsley.

Schafer, Kristin. 2004. "Biomonitoring: A Tool Whose Time Has Come." At http://www.panna.org/legacy/gpc/gpc-200404.14.1.02.dv.html.

———. 2006. "One More Failed U.S. Environmental Policy." Foreign Policy in Focus September 6. At http://www.fpif.org/fpifxt/3492.

Scheper-Hughes, Nancy. 1992. *Death without Weeping: The Violence of Everyday Life in Brazil.* Berkeley: University of California Press.

Scheper-Hughes, Nancy, and Carolyn Sargent, eds. 1998. *Small Wars: The Cultural Politics of Childhood.* Berkeley: University of California Press.

Schettler, Ted, Gina Solomon, Maria Valenti, and Annette Huddle. 1999. *Generations at Risk: Reproductive Health and the Environment.* Cambridge, Mass.: MIT Press.

Sedgwick, Eve Kosofsky. 1991. "How to Bring Your Kids Up Gay." *Social Text,* 29: 18–27.

Selevan, Sherry G., Libor Borkovec, Valerie L. Slott, Zdena Zudova, Jiri Rubes, Donald P. Evenson, and Sally D. Perreault. 2000. "Semen Quality and Reproductive Health of Young Czech Men Exposed to Seasonal Air Pollution." *Environmental Health Perspectives* 108(9): 887–94.

Sexton, Ken, Larry L. Needham, and James L. Pirkle. 2003. "Human Biomonitoring of Environmental Chemicals: Measuring Chemicals in Human Tissues Is the 'Gold Standard' for Assessing People's Exposure to Pollution." *American Scientist* 92(1): 38–45.

Shildrick, Margrit. 1997. *Leaky Bodies and Boundaries: Feminism, Postmodernism, and (Bio)Ethics.* London: Routledge.

Smolin, David M. 2006. "Overcoming Religious Objections to the Convention on the Rights of the Child." *Emory International Law Review* 20: 81–110.

Sokol, Rebecca Z., Peter Kraft, Ian M. Fowler, Rizvan Mamet, Elizabeth Kim, and Kiros T. Berhane. 2006. "Exposure to Environmental Ozone Alters Semen Quality." *Environmental Health Perspectives* 114(3): 360–65.

Specter, Michael. 2002. "The Long Ride." *New Yorker,* July 15.

Stafford, Barbara Marie. 1991. *Body Criticism: Imaging the Unseen in Enlightenment Art and Medicine.* Cambridge, Mass.: MIT Press.

Staiger, Janet. 2005. *Media Reception Studies.* New York: New York University Press.

Star, Susan Leigh. 1991. "The Sociology of the Invisible: The Primacy of Work in the Writings of Anselm Strauss." In *Social Organization and Social Process: Essays in Honor of Anselm Strauss,* ed. David Maines (265–83). Hawthorne, N.Y.: Aldine de Gruyter.

Star, Susan Leigh, and Anselm Strauss. 1999. "Layers of Silence, Arenas of Voice: The Ecology of Visible and Invisible Work." *Computer Supported Cooperative Work* 8: 9–30.

Stein, Rachel, ed. 2004. *New Perspectives on Environmental Justice: Gender, Sexuality, and Activism*. New Brunswick, N.J.: Rutgers University Press.

Steingraber, Sandra. 2001. *Having Faith: An Ecologist's Journey to Motherhood*. New York: Perseus.

Stokstad, Erik. 2004. "Pollution Gets Personal." *Science* 304(5679): 1892–94.

Sugg, Diana. 2002. "Cruelest Mystery: Death before Life." *Baltimore Sun*, March 24.

Svedin, Carl Goran, Kristina Back, and Radda Barnen. 1997. *Children Who Don't Speak Out: About Children Being Abused in Child Pornography*. Stockholm: Radda Barnen (Swedish Save the Children).

Swofford, Anthony. 2003. *Jarhead: A Marine's Chronicle of the Gulf War*. New York: Scribner.

Takacs, Stacy. 2005. "Jessica Lynch and the Regeneration of American Identity and Power Post 9/11." *Feminist Media Studies* 5(3): 297–310.

Talley, Heather Laine, and Monica J. Casper. Forthcoming. "Oprah Goes to Africa: Philanthropic Tourism, Celebrity Culture, and the Politics of (Dis) Engagement." In *The Story of O: An Edited Volume on Oprah L. Winfrey*, ed. Kimberly Springer and Trystan Cotton. Jackson: University Press of Mississippi.

Terrazzano, Lauren. 2006. "Armstrong an Inspiration to All with Cancer." *Newsday*, November 6, D9.

Thomas, W. I., and Dorothy Swaine Thomas. 1970. "Situations Defined as Real Are Real in Their Consequences." In *Social Psychology through Symbolic Interaction*, ed. Gregory Prentice Stone and Harvey A. Farberman (154–55). Waltham, Mass.: Xerox Publishers.

Thompson, Christopher S. 2006. *The Tour de France: A Cultural History*. Berkeley: University of California Press.

Treichler, Paula. 1999. *How to Have a Theory in an Epidemic: Cultural Chronicles of AIDS*. Durham, N.C.: Duke University Press.

Tucker, Bruce, and Priscilla Walton. 2006. "From General's Daughter to Coal Miner's Daughter: Spinning and Counter-Spinning Jessica Lynch." *Canadian Review of American Studies* 36(3): 311–30.

Turner, Bryan. 1987. *The Body and Society: Explorations in Social Theory*. Oxford: Blackwell.

———. 1992. *Regulating Bodies: Essays in Medical Sociology*. London: Routledge.

UNAIDS. 2006. "The Impact of AIDS on People and Societies." In *2006 Report on the Global AIDS Epidemic*, chapter 4. At http://data.unaids.org/pub/Global Report/2006/2006_GR_H04_en.pdf

UNDP. 2005. *Human Development Report 2005*. At http://hdr.undp.org/en/reports/global/hdr2005/.

U.S. Census Bureau. 2006. Fact Sheet. American Community Survey, Memphis, Tenn.

U.S. Department of Defense. 2006. Defense Manpower Data Center. Unpublished data, September 30. At http://www.dmdc.osd.mil.

U.S. Department of Justice. 2000. Sexual Assault of Young Children as Reported by Law Enforcement: Victim, Incident. Washington, D.C.: Bureau of Justice Statistics.

"US Soldier Admits Murdering Girl." *BBC News*, February 22. At http://news.bbc.co.uk/2/hi/americas/6384781.stm.

Vastag, Brian. 1999. "High-Profile Cancer Cases Prompt Awareness Efforts." *Journal of the National Cancer Institute* 91(21): 1802.

Vesely, Rebecca. 2003. "Study: Women Bear Brunt of Environmental Toxins." *Women's E-News,* October 21.

Waldman, Paul. 2004. *Fraud: The Strategy behind the Bush Lies and Why the Media Didn't Tell You.* Naperville, Ill.: Sourcebooks.

Walker, Rob. 2004. "Yellow Fever: What Happens When Philanthropy and Style Team Up? A Charity Craze." *New York Times,* August 29, 23.

Wang, Richard Y., Michael N. Bates, Daniel A. Goldstein, Suzanne G. Haynes, Karen D. Hench, Ruth A. Lawrence, Ian M. Paul, and Zhengmin Qian. 2005. "Human Milk Research for Answering Questions about Human Health." *Journal of Toxicology and Environmental Health, Part A* 68: 1771–1801.

Washburn, Rachel. Forthcoming. "Measuring Chemicals, Making Bodies: Mapping the Social Terrain of Human Biomonitoring in the United States." Ph.D. diss., University of California, San Francisco.

White, Rob. 2002. "Social and Political Aspects of Men's Health." *Health* 6(3): 267–85.

Whitelaw, Kevin. 2007. "Selling a Convenient Truth." *U.S. News and World Report,* May 7, 39.

Williams, Florence. 2005. "Toxic Breast Milk?" *New York Times,* January 9.

Williams, Joe. 2005. Review of *March of the Penguins. St. Louise Post-Dispatch,* July 8.

Williams, Kayla, with Michael E. Staub. 2005. *Love My Rifle More Than You: Young and Female in the U.S. Army.* New York: Norton.

Wilson, Jan Doolittle. 2007. *The Women's Joint Congressional Committee and the Politics of Maternalism, 1920–30.* Urbana: University of Illinois Press.

World Bank. 2005. "Momentum to Fight AIDS Gains Pace in Moscow." *News and Broadcast,* April 4. At http://web.worldbank.org.

Wright, Kai. 2005. "Super Infector." *ColorLines* 31(Winter): 1–4. At http://www.colorlines.com/article.php?ID=34.

Yalom, Marilyn. 1998. *A History of the Breast.* New York: Ballantine.

Zacharek, Stephanie. 2005. "March of the Penguins." *Salon.com,* June 24.

Zoepf, Katherine. 2007. "Desperate Iraqi Refugees Turn to Sex Trade in Syria." *New York Times,* May 29.

Index

About the Authors

MONICA J. CASPER is Professor of Social and Behavioral Sciences and Women's Studies and Director of Humanities, Arts, and Cultural Studies at Arizona State University's New College of Interdisciplinary Arts and Sciences. She is author of *The Making of the Unborn Patient: A Social Anatomy of Fetal Surgery* and editor of *Synthetic Planet: Chemical Politics and the Hazards of Modern Life.*

LISA JEAN MOORE is Professor of Sociology and Women's Studies and Coordinator of Gender Studies at Purchase College, SUNY. She is author of *Sperm Counts: Overcome by Man's Most Precious Fluid* (NYU Press) and coauthor of several articles and books focused on the body.